Goo Bless!

LORD
OF THE
SKIES

D0907335

LORD
OF THE
SKIES

**EXPLORE THE MIRACLE OF WEATHER,
AND REDISCOVER GOD, HIS SON,
AND YOURSELF**

ALAN S. WINFIELD

WINEPRESS PUBLISHING

ISBN 1-57921-233-6
Library of Congress Catalog Card Number: 99-63732

DEDICATION

To my wife, Pam, who like the Lord, and despite
myself, will never forsake me.

CONTENTS

ACKNOWLEDGEMENTS

I don't believe I could ever have gone through the ordeal of writing a book without a host of people along the way who have encouraged my ideas. It also has helped to have a target audience in mind. The following folks have my eternal appreciation for the roles that they have played in my life and in the writing of this book.

My mother is probably the biggest shaper of who I am today and also my biggest supporter. The love, self-worth, and encouragement that one derives from God is infinite, but the love, self-worth, and encouragement that my mother passes my way, comes pretty close. I don't believe I'll ever know another person in my life who believes in me as much as my mother does. I love her dearly.

In my childhood I had a very special, very Jewish, and very colorful Aunt Irva, who always encouraged my desire to be a meteorologist. Despite other family members' attempts to channel me into other professions with a more promising and prestigious future, she would argue to let me be what I want to be. I love her for standing by me and for seeing something inside me.

I thank God for putting three key men in my life who would ultimately be my models of what Christian men are really about. They took time to plant seeds in me, despite my immature soil. They are Scott Whitmore, Michael Carter, and Russell Kinsaul. All three are the greatest men of integrity I have ever known.

I am grateful to a group of folks at West End Assembly of God Church in Richmond, Virginia, who let me lead their church Bible study group. It was in that forum that I was first able to speak about the concepts in this book. They graciously let me speak my heart and mind.

To this day, the original leaders of that study group, Ernie and Linda Campe, have my unwavering gratitude. They are my heroes because of the call to missions that they pursued. They are also my fond friends, and have been steadfast encouragers as I embarked upon this book.

Mike Corbin, a Christian co-worker from Richmond, also has my gratitude. Many a time he and I would go out to lunch, and I would chat away about my dream of writing this book. He politely let me make him my sounding board. He was my Christian friend, and boy, could he make me laugh.

I am indebted to the Puyallup Church of the Nazarene in Puyallup, Washington. Dr. Leslie Parrott II, Mark Dennis, and Tim and Carole Stratford were instrumental in pushing my walk with the Lord to a new level. They also guided me as I initiated the pages of this book. In my dictionary, these people define "quality."

Over the years I have had the privilege of having a terrific friend, Kevin Selle. He and I have shared with each other the ups and downs of life's cycles. He has been a staunch encourager as this book has unfolded. I am grateful to him and his wife.

I also want to give a special thank you to Thom Savela who helped me create Figures 1, 3, 5, 6, and 7.

In the first paragraph, I mentioned a target audience. That audience, for whom I am writing this book, is my true motivator. Part of that audience is all the people who are unsaved and need to hear the good news about God and His Son, Jesus Christ. However, this book is just as much intended to encourage Christian believers. While writing this book, it has helped to put a face to this large group of people. The face that carried me through was that of my seven-year old son, Austin. I want him to know that I am a scientist, a man, a husband, and most importantly, his daddy, who has chosen to make my walk with the Lord the number one thing in my life. I want Austin to know that this book is what I stand for and who his earthly father is. My hope is that I will have somehow helped him to have a better understanding of who his Heavenly Father is.

INTRODUCTION

It was an evening exam. Those are the worst. It's true you have all day to study, but you also have all day to worry. I couldn't even "enjoy" one last mystery-meat dinner in the Penn State dorm cafeteria. Finally, 7 P.M. arrived, the start time for my very last final exam. As I walked into the test room, displacing an exiting crowd of drained and relieved students from the prior exam period, nostalgia overtook me. After this exam, I would never be in this room again. After this exam, I would not see most of these people again. After this exam, I would be graduating and receiving my diploma. After this exam, I would be a real meteorologist. After this exam, I would never have to memorize-to-pass an exam again.

Now don't pretend that you don't know what I'm talking about when I say memorize-to-pass. It's a skill everyone learns quickly in order to survive at college. There's no time to learn calculus and physics and computer science and anthropology and geology and geography and French and chemistry and vector analysis and thermodynamics and English literature and statistics, let alone meteorology. So instead, you shove in your brain all it's going to take to pass each test.

I've always had a passion for meteorology. While most folks I knew, even in college, were still trying to figure out what they wanted to be, I had the luxury of having a burning desire to be a meteorologist since kindergarten. Growing up, I bypassed the weekly whims of first choosing to be a fireman, then a policeman, then a doctor, and then the president.

My interest in meteorology stemmed from a strong dislike of school. At the wee age of five, growing up in Convent Station, New Jersey, I learned very quickly my first mathematical equation: *snow equals no school.* I tuned in to the weather in hopes of hearing that magic four-letter word, snow. That's all it took to hook me on meteorology.

Unbeknownst to me at the time, I was nurturing a call to meteorology. The family telephone quickly became the nightly pipeline connecting my class with me. Classmates would call wanting to know if they needed to do homework, or if instead snow would be canceling school. Weather was certainly something to be excited about. Eventually, I began branching out, as all inclement weather, not just snow, was turning me on. While dad cursed every storm that visited our fair town, I was doing a celebration dance right in front of him. His most repeated lines to me were, "You only think you love this weather. Wait until you're older and have to deal with it." Dad, I'm older now, but my wife will vouch for me that I'm still doing that dance of excitement with every storm.

The more folks hated the weather, the more I loved it. It got tough conversing with them, though. In fact, confusion easily ensued when I would talk about the great weather headed our way. My great weather was their lousy weather. Even now, as a meteorologist on television, I have to watch myself, or my excitement and joy over a storm will show through. I basically lie on-air, since I have to make people think I'm with them on the belief that sunshine is good weather and all else is bad. My nose is getting long.

My dislike for school never changed. However, my love of weather overshadowed that so I went on to pursue a graduate degree at the University of Oklahoma. My memorize-to-pass method got me through until the last semester. That is when I had the final hurdle to jump over. That hurdle seemingly extended from the stratosphere down to the ground. The hurdle was the comprehensive exams. My professors all said, "Passing is simple as long as you really know it all." I started preparing well in advance and quickly realized I could not memorize everything. It was too much. Imagine that, I'd really have to learn! These people had nerve. My brain cells quivered, leading the rest of my body to basically convulse. This was one ailment I couldn't blame on all those years of dorm food.

As I set out to conquer the mother of all exams, I chose to truly prepare by learning. In the process, something wonderful happened. The information began to fit together like a Milton Bradley jigsaw puzzle. I was genuinely absorbing and learning. Instead of tons of disjointed facts and formulas, I found one concept naturally flowed and led to another. I was on my way!

Looking back, I wondered how I could have memorized all of those years. I was cheating myself out of the thrill of discovery. Weather was a beautiful and awesome system. It was perfect. Everything was interconnected and nothing was disjointed. It was a system surrounding the earth that was worth beholding. As I understood more and more, the task of learning stopped being a task at all. Equations and concepts became obvious, based on the beauty of the way weather is set up. I was even able to come up with my very own Seven Laws of Meteorology. These were personally discovered categories into which all facets of weather could fall. I watched others struggle to prepare for these exams. I was enlightened and sparked, and they were depressed.

My master's degree was completed, and I went back to Penn State to get the final academic prize, a Ph.D. A short time later I became a Ph.D. dropout. Oh, the shame I brought to my family. They'd never had a dropout before. Basically, I was chomping at the bit to go to work in weather. You see, as I stated earlier, my dislike for school never changed. However, my love of weather overshadowed that.

After a stint at being a meteorology instructor, I got called into television. The job site: Lawton, Oklahoma. I'm telling you, as a meteorologist, I had it all. I was smack-dab in tornado alley. I had the Doppler radar on one side of me, and reams of weather data and computer graphics on the other. It was life at its best.

Lawton lay in the shadows of the Wichita Mountains. The folks back east envisioned Oklahoma as a flat, dusty, rattlesnake infested countryside. I encouraged their stereotype with a story that there was talk about some roads actually being paved within the year.

It was a paved road though that took me to a jewel of a location, in the heart of the mountains. There, sitting on the edge of a pristine lake, I often became lost in its reflection of the rugged mountains flanking me all around. The mountains were a unique landform, basically heaps of boulders that mounded majestically upward in concert. Wildflowers and cactus graced the hills. On some days, a small herd of buffalo would wander by. The beauty, though, did not stop at ground level.

Gazing up from this spot, the sky was like a theater. Many times I witnessed the metamorphosis of an ocean blue sky into a boiling cauldron of water vapor. Invisible heated plumes of air would rise off the landscape, transforming into visible clouds that erupted in all directions like a volcano. Soon the fireworks would begin, as trees of lightning would connect the sky to the ground. Waves of thunder would then roll across the landscape. My ears and eyes were being tantalized, but it was an unspoken sense within me—a sense that we all have—that was awakened and activated. God was before me. My mind quickly began a thoughtful journey, as weather became my avenue to God.

In the chapters ahead, I will take you on this journey through system weather. It is a journey through one of God's most magnificent creations. Along the way, we will explore my Seven Laws of Meteorology. They are seven encompassing laws that describe all of weather. We will then discover that these seven laws are legislated in the earliest chapters of the Bible. They are thus God's laws, and not mine. They are so encompassing, that we will come to recognize that these laws don't apply to just weather, but to the rest of God's creations as well. Most importantly, they apply to us, and the life we can have centered on a savior named Jesus Christ.

It took a thunderstorm in the Wichita Mountains for the awareness of God's presence to click with me. I have since come to realize that God has always been, and always will be, with every one of us. Some of us just don't realize it yet. Unfortunately, some never will. Nonetheless, that point in time in Oklahoma was my life-changing moment. The realization I suddenly had of God overwhelmed me in a very memorable way.

In fact, when that realization of God struck, I might as well have been hit by lightning. I became *charged* with excitement. I became *electrified* with awe. I also became *stunned* by the fact that I had been blind until that moment.

My vision has since improved considerably. I hate clichés, but darned if they don't come in handy, so I'm going to use one: the world is a stage. Personally, the stage curtain just keeps rising for me as my sight matures. Initially, my sight was focused on a thunderstorm amidst the Wichita Mountains. All I saw was system weather. Obviously, though, other systems do exist. Now I see that all these systems are part of a much larger, interconnected, and even more ingenious, intricate, and miraculously grand system, designed by God. His grand system contains everything He ever created in the world and universe that surround us.

If you think I was thrilled before, you should see me now. I'm absolutely high! What I used to take for granted I now see in a whole new light. God is revealed in every system that can be contemplated. Furthermore, His systems are of brilliant design.

These systems before us are a pathway to God. The Bible sums this up in Romans 1:20 (NIV), "For since the creation of the world His invisible attributes, His eternal power and divine nature, have been clearly seen, being understood through what has been made, so that they are without excuse."

I would be without excuse if I didn't share with you the enlightenment I have about the brilliance, intricacy, and miracle of system weather. As a scientist, it is my conclusion that it is impossible to dissect the science of meteorology without involving God. Just as music would be chaos without a beat for the notes to organize around, meteorology would be inexplicable without a centering of it around the Almighty Creator! My enlightenment centers on what I call the Seven Laws of Meteorology. They are indeed a pathway to God.

ONE

The Weather is Law-Abiding

A man with a funny tasseled hat, a medal-adorned robe, and an air of royalty greeted me and escorted me to the microphoned podium. His name was Larry, so I called him Larry, but the local community service club members referred to him much differently. My memory is fuzzy, but I believe they called him something like "your exalted royal highness and king grand Pooh-Bah." If you ask me, I still think he looked much more like a Larry. Larry's introduction of me was shorter than his title, and so I quickly began my weathertalk.

As a television station employee, I am often invited to schools, civic organizations, and the like to deliver weathertalks. These talks are basically a discussion on meteorology, and they serve as a public relations event for the station. Normally, when I do weathertalks at civic organizations, the crowd is quite receptive. I typically begin by stating that the sun is the origin of all our weather, and I share with them how weather is spawned. Next, I tell them about the nature of my job, the tools I use, humorous things that have happened, and so on. The talk ultimately turns into a fun question-and-answer session, which is my favorite part. It is a chance to cater my weathertalk to exactly what the crowd is interested in. This format has successfully repeated itself wherever I have gone, at least until this particular day.

Yes, on this day, sitting in the auditorium's fourth row, hidden in the slight shadows of two men on either side of him, sat a short, stocky, club member. He had obviously ordained himself to be "the grand exalted obnoxious one." It seemed that no matter what I said, he felt he had to interrupt with a sarcastic remark. If I talked about my job as a television meteorologist in Seattle, he would shout out that I have the only job in the world where I can be wrong every day, but still get paid. If I talked about the computers I use to make forecasts, he would exclaim he could tell more about the weather for the next few days just by looking out the window once a week. If I would talk about our expensive Doppler radar and satellites, he would bellow out that I should save my money, since the Washington state forecast was never anything but rain and more rain.

At first my game plan was to politely play off his remarks. Unfortunately, that only encouraged him. Finally, I altered my comfortable format, and the result amazed me. Boldly, I told the group that it was obvious some people just can't handle a meteorologist in their midst, but that's fine because I get my self-worth from God above. For the first time in thirty minutes, not a peep was heard from my "friend."

Staring intently at the troublesome "gentleman," I proceeded to launch into an impromptu discussion about seven laws of weather that had been set forth in the Bible. These were laws of God that I considered every time I made a forecast. I then explained how these laws could apply to everything around us including the way we conduct our lives. When my time was up, the thundering applause gave me goose bumps. I mentally redirected the applause upward.

I want to get goose bumps again as I share with you the Seven Laws of Meteorology that I came up with years ago during preparations for my graduate school exams. The discovery I made later, about my Seven Laws of Meteorology actually being from God, at first startled me, then thrilled me, and then saved me. The laws are actually a design of God's that is clearly revealed in Genesis, the very first chapter of the Bible. The laws are then upheld through the remainder of the Bible.

Before we journey through the Bible, though, we must first outline the seven laws. In actuality, these laws are seven all-encompassing categories into which all aspects of weather fall:

Law 1: The sun takes center stage with weather.

Law 2: Weather is a total system.

Law 3: Weather smoothes out excesses and deficits.

Law 4: An attitude of 'whatever it takes' is behind the design of weather.

Law 5: Balance is a characteristic of weather.

Law 6: All weather serves a purpose.

Law 7: Weather contributes to entropy on earth.

Another way to appreciate the above list is to put it into a picture. This allows us to view the way that the seven laws are connected to each other.

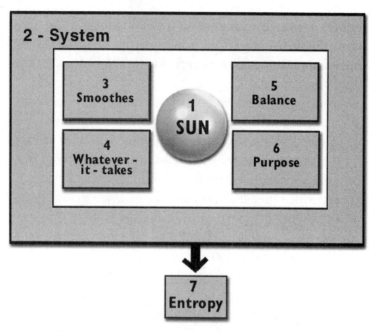

Fig. 1. Configuration of the Seven Laws of Meteorology

In the figure above, the outer rectangle represents the total system of weather. It is Law 2, and is labeled with the number two. Inside this rectangle is a centralized circle, signifying the sun, upon which all weather is centered. This symbolizes Law 1. Evenly distributed around the sun, but within the confines of the total system, are four governing characteristics of weather. These are Law 3 through Law 6, which stand for smoothing, a whatever-it-takes attitude, balance, and purpose. Finally, the entire system of weather leads to a degradation of the planet. This degradation is known as entropy. The downward arrow to the number seven depicts Law 7.

In the next seven chapters, we will explore each of the seven laws in chronological order. Each chapter will have a fairly standardized format. First, each chapter will focus on the "Meteorology of a Law," directly followed by the "Biblical Source of that Law." Next, the chapter will contain a section that I refer to as the "Biblical Expansion of the Law." This refers to Biblical applications of the law beyond the realm of just weather. In fact, this will demonstrate how far-reaching Law 1 through Law 7 really are. Finally, every chapter will conclude with a section called "Pathway to Jesus Christ." This portion of the chapter will be devoted to discovering how the individual law applies to our lives, as well as the way it can lead us to a personal relationship with Jesus Christ.

TWO
●●●●●●●●●●●●

Law 1: The Sun Takes Center Stage with Weather

METEOROLOGY OF LAW 1

It was the summer of 1971, and the neighbors across the street had just bought a splotchy black-and-white cat. I was eleven years old, and I remember thinking that if it were a dog, it would have been called a mutt. As a cat though, they affectionately called it Ra, and they adored that feline. It ate the choicest cat food money could buy. It was allowed to roam inside the house and outside, depending upon its whim at the moment. Ra had his own little bed to sleep in and more toys than my seven-year old will ever have.

That was also the same year I was taking a history class in school. Of course that was more than a couple of decades ago, so I suppose that makes it qualified to now be called "ancient" history. One of the few things I recall from my class was the amazing number of gods that ancient civilizations worshipped. In particular, the Egyptians had a sun god they worshipped called Ra. If the

Egyptians worshipped their god Ra anything like the folks across the street worshipped their cat Ra, it must have been quite a sight.

Other civilizations have had their share of sun gods as well. Usually these sun gods were pretty much on the top of their "god-totem-poles". The Romans had Apollo, the Greeks had Phoebus Apollo, the Incas had Inti, the Mayans had Hun-Ahpu, the Hindus had Surya and Mitra, and so on. These ancient peoples all realized that weather, and thus life itself, were only possible with the sun's existence. As a result, they chose to worship the sun. (Heck, back in my younger days you'd even find me laying out, catching a few rays, and calling myself a sun worshiper.)

All of the ancient sun gods have come and gone, and the threat of skin cancer has ended my sun worshipping, but the truism that remains is that the sun is a miraculous entity. The sun itself is not a god, but there is shining brilliance in the great plan to center our very existence on it. The sun is in fact the center of system weather and thus the center of all life on earth. It is earth's perpetual furnace.

The sun itself is not a god, but there is shining brilliance in the great plan to center our very existence on it.

By design, the ball of fire that graces the sky daily is a monstrous mass of fiery energy. Since energy is directly proportional to temperature, it's interesting to note that the sun's temperature exceeds ten thousand degrees Fahrenheit. That means it has tons of energy. Part of its energy rockets toward the earth and gives us light. Part of its energy rockets toward the earth and gives us warmth. Part of its energy rockets toward the earth and feeds and grows vegetation. Part of its energy rockets toward the earth and interacts with plants, generating oxygen for all living things. Part of its energy rockets toward the earth and can be collected for energy purposes. What impresses me most, though, is the part of its energy that rockets toward the earth and generates all of our weather.

This may come as a shock to some people, but the atmosphere is a liquid. When solar energy shoots into the earth's atmosphere, the atmosphere reacts in much the same way as a pitcher of sun-tea reacts when it is placed outdoors to brew. As the sun's energy is imparted into the atmosphere, warm currents of air begin to rise vertically and travel horizontally. Again, this is strikingly similar to a pitcher of sun-brewed tea, because if you look carefully at the fluid in the tea pitcher, you can see the tea-stained water rising and swirling. It is this same sun-induced swirling and rising in the atmosphere that leads to cloud formation. It also results in moving air masses, storms, rain, wind, snow, hail, sleet, freezing rain, lightning, thunder, tornadoes, and hurricanes. The sun is the source of our weather.

The sun is truly the powerhouse behind the earth's weather machine. The sun literally gets the atmosphere moving. If the sun didn't exist, weather would not exist as well. That is why the sun takes center stage in the seven-law configuration of weather.

BIBLICAL SOURCE OF LAW 1

I used to believe I was quite the architect back in my college days, when I designed my own seven-law blueprint for weather, placing the sun in the center, as Law 1. Of course, I now know it was the great architect, God, who indeed designed weather with the sun at center stage. This became very apparent to me as I read the Bible.

There's a man named Ken Ham who reads the Bible a lot. He is the executive director of a Bible-based ministry called Answers in Genesis. To quote the ministry's mission, it is "a Christ-centered, evangelistic ministry dedicated to defending Scripture from the very first verse." Ken Ham uses the verses of Genesis in many ways, including to defend creationism versus evolution, to talk about dinosaurs, to get a handle on the great flood, and to discuss the big bang. I believe I am going to make Ken happy, because it is also from the very beginnings of the Bible, in Genesis, that I find the roots of Law 1 to be God's.

In fact, it is in Genesis 1:3 that God begins to set forth Law 1. "And God said, 'Let there be light,' and there was light." This is the passage where He establishes light over the dark earth. We don't quite have the sun yet, but we are on our way. What's worth noting is that after making and describing the heavens and the dark earth in Genesis 1:1-2, creating light became God's very first order of business. In fact, it was only after light was established that everything else could follow.

God was so happy with light, that in Genesis 1:16, "God made two great lights. The greater light [the sun] to govern the day and the lesser light [the moon] to govern the night." Now we have the sun. Interestingly enough, moonlight is in reality reflected sunlight, so the sun in essence governs all of the time!

BIBLICAL EXPANSION OF LAW 1

God created the sun as a tool to give us light and weather here on earth. By it being the very source and generator of our weather, we witness one of the ways that the sun governs and sustains our lives. The miracle of the sun, though, definitely extends far beyond the realm of weather.

The sun is actually the center of many other life-maintaining systems. The sun accomplishes this in several ways. First, the sunshine is a food source for plants. These plants are in turn a food source for man and animal alike. So from the sun comes food that is vital to our existence. Second, the sun maintains and protects life by interacting with plants and creating the air we breathe. This is done via the plant process of photosynthesis. So from the sun comes air, which is also vital to our existence.

The sun is a true miracle. It is by far one of God's greatest tools for maintaining life on earth. Again, I exclaim, what brilliant architectural design. Law 1 states that the sun takes center stage with weather. It should probably be expanded to say the sun takes center stage for life on earth.

Moses was one who recognized the key role the sun played in the lives of God's children. In fact, during a blessing he gave to the tribe of Joseph, found in Deuteronomy 33:13-14, Moses said, "May the Lord bless his land...with the best the sun brings forth...." Moses knew the crucial effect the sun could have on a man's land, his livelihood, and thus his life.

In Psalm 84:11, the power and scope of the sun's task is demonstrated when the writer of the Psalm states that "the Lord God is a sun and shield." What an interesting way to describe our Heavenly Father. This is the writer's way, no doubt, of describing the magnitude of God's ability to protect us like a shield and to sustain and bless us like the sun.

Unfortunately, many civilizations in the past were so enthralled by the sun, that they chose to worship it rather than God, who created it. In His infinite wisdom, God foresaw that this would happen. He even warned against it through Moses, who speaks in Deuteronomy 4:19: "And when you look up to the sky and see the sun, the moon and the stars—all the heavenly array—do not be enticed into bowing down to them and worshiping things the Lord your God has apportioned to all the nations under heaven." As history shows, though, not all civilizations have heeded this warning.

LAW 1: PATHWAY TO JESUS CHRIST

God created the sun to govern the sustaining of life on earth. After all, because of the sun, plants grow and yield food. Because of the sun, weather is driven, the rains fall, and thirst is quenched. Because of the sun, vegetation flourishes, animals thrive, excess energy is dispersed and so on. He centered our planet on the s-u-n and miraculously set up everything just right to sustain us.

God is definitely in the habit of centering. In fact, He has also given us his Son—s-o-n—Jesus Christ, to be the center of our lives. In the same way that the sun brings life to earth, Jesus Christ brings life to our very being. Because of the Son, our thirst is quenched, we are fed, we flourish, we thrive, our fears are dispersed, and so on.

The above thought is captured so eloquently in the movie, *Amistad*. One particular scene shows an unclear future for a group of Africans who have been kidnapped from their homeland and taken to America to be slaves. They have been abused and tortured and are now suffering miserably while being held in a U.S. prison. They only speak their native tongue and have no clue what is happening to them as they await a trial. In the midst of their horror, one of the Africans obtains an English Bible, and is at least able to understand the pictures in it. He is able to accurately recreate the story of Christ for himself as he chronologically shuffles through the pictures, as if he were playing connect the dots. A calming peace now surrounds him.

He later shows a fellow African the pictures. In his own language, he points out the glaring similarity between their lives and the life of the man in the pictures, who happens to be Christ. There in the book was a man just like them, who at one time was free and suddenly was tortured and thrown into prison. The final Bible pictures showed that all ended well for this man, as he rose triumphantly up toward the sky. That gave the African solace.

Now here is the line, in the movie, that I most cherished. The African pointed at the pictures of Jesus and said, "See how the sun follows Him." Sure enough, in every picture in that old Bible, just as in Figure 2 below, Christ was depicted with his head overlaid on a sun. This set Him apart from the ordinary people that surrounded Him. The African recognized that this man with the sun was special. After all, Christ was able to overcome and conquer whatever circumstances came His way, even death!

2. Rembrandt's painting,
The little children being brought to Jesus
Used by permission, Rijksmuseum, Amsterdam

From the sun to the Son. I'm not surprised both went hand in hand in that Bible picture. Both are items that our lives can be centered around. It is involuntary on our part that our lives are centered on the sun. God created it that way. It is completely voluntary though, whether we choose to center our lives on the other Son, Jesus Christ. Again, God created it that way.

As you look around, it's interesting to note what other items nonbelievers often center their lives upon. Sometimes it's superstitions. For others it's centering their lives on possessions. Some choose work to be the focus of their lives. Drugs and alcohol are options also chosen. Some even choose other people to center their lives around.

I once worked with a woman who depended totally on her horoscope. She was a devout believer and follower of the local newspaper's daily horoscope column. It was the first thing she checked every morning. (By the way, that's as bad as putting faith in the Farmer's Almanac's daily weather forecasts!) From time to time she would try to share her horoscope with me while at work in our television newsroom. One particular time I was ready for her. Her newspaper said she should "avoid monetary dealings for the next 24 hours." I then grabbed another city's newspaper since the station subscribed to many. The horoscope in that paper, obviously written by a different columnist, told her "financial investments are highly worth the risk. Seek fortune today!" Uh oh. We now had a contradiction between horoscope columns!

My co-worker went speechless. Her face paled. I had definitely burst her bubble. I kid you not when I tell you this was a very serious matter for her. She looked as if someone had just jolted her with a stun gun. Her life had just been shoved off center.

Invariably, when your life is centered on anything or anybody other than Jesus Christ, you become a hostage to it. You become a prisoner to its precarious nature. Your life's centering, unless on Christ, will only roll with the tide. It is unstable and easily overturned. Life comes down to a choice for all of us. A choice between holding firm and steady to the rock of Christ, or being held hostage by a foundation composed of shifting and sinking sand.

The option to end the hostage situation that you may be in right now is available to everyone. This option involves taking part in a two-way relationship between you and Jesus Christ. Your situation may appear hopeless, but the rock of Christ will pummel that hopelessness. The result will be a new centering of your life around the Son of God. He is a rock that is impossible to move or overturn.

> **Invariably, when your life is centered on anything or anybody other than Jesus Christ, you become a hostage to it. You become a prisoner to its precarious nature.**

The words of Christ are a blueprint for the salvation of our lives. I'm not just talking eternal salvation, which allows us to enter the kingdom of heaven once our physical bodies die. There is another salvation that begins right now. It is the salvation of our earthly lives. We experience a return of purpose, hope, worth, and internal peace to our lives. Christ's ways reflect God's purpose for us, so Christ's ways are worthy of being mirrored.

I promise you that if you want to experience a stable, dependable, caring, and loving person to center your life upon, there is but one choice. That is Jesus Christ. All other options you think you have will only let you down.

I also promise you that if you want to fully experience God in your life, all you need do is take advantage of His loving offer to center your life around His Son. Simply invite Him into your heart today. Let people say about you, "See how the Son follows him."

THREE

Law 2: Weather is a Total System

METEOROLOGY OF LAW 2

I t had been a long, exhausting workday. The commuters had now made it home. In many instances, they had already finished their dinners and were settling down to playtime with the kids or maybe some relaxation in front of the television. It was an evening that had all the signs of fading and being forgotten over time until the landscape decided to lurch and roll and shake. An earthquake had pierced the normal routine of the night.

Coverage began immediately on the local television stations. Two stations broke in and revealed that they had a disaster plan that was being implemented. All of their employees knew the plan and had already been dispatched to select locations. They had a system in place, and it was impressively unfolding on the air. Within moments, the situation was covered from every conceivable angle.

My television station did not fare as well, however. The folks in the newsroom were in a panic. The phones were ringing endlessly, everyone was shouting at once, there was a paralysis about what should be done first, no plan had ever been developed ahead of time, the team leader never arrived, and the on-air product became an attempt to play catch up with the other stations. You see, no system had ever been designed.

I took Management 101, so watch out, I'm dangerous. One of the first items in the curriculum is to have a system or plan ready, to deal with the inevitable unexpected. If you have a system in place, the product does not fall apart. If you don't have a plan, then there's nothing to hold the product together when adversity strikes.

Fortunately for the world, there is a firmly established system of weather that is fixed and can never fall apart. It is a closed system that is totally interconnected. It is elaborate and beautiful. It has rules that it obeys. It can even be described by sets of equations.

Did I say equations? I always wondered why I had to learn so much math. It's kind of funny. None of my teachers ever revealed what the purpose of all of the math was. Year after year I crunched numbers in bizarre-looking formulas, but never knew why. What did differential equations have to do with me? How could calculus affect the world? How did the eigenvector of a linear operator ever change anyone's life?

When I had finished all of my math courses and began the meteorology, mathematical enlightenment finally hit me. After fifteen years of math, I was now seeing the first applications. It was an eye-opener. It was a new world out there. Life was good again. All right, I wasn't having a religious experience, but I was certainly having a math revival.

It turns out that everything around us can be described in two ways. One way is to describe it using the everyday language that we speak. The other way is to use expressions in the language of mathematics. That means I have the choice of describing everything mathematically, or in English.

Let's use an example of a developing cloud. As a kid, we all remember lying on our backs staring up at the sky, making animal pictures out of the clouds. Even today, you may note that a particular cloud has evolved into a shape that looks like a duck, but you can also describe the growth of the mass of cloud droplets by the equation:

$$J = 4\pi r_c^2 \frac{e}{\sqrt{2\pi mkT}} Z_n \exp\left(-\frac{4\pi r_c^2 \sigma}{3kT}\right)$$

By describing any process, thing, or phenomenon with an equation, you have the advantage of being able to mathematically play around with it. In the above equation, you can actually determine what happens to the cloud's shape and growth rate by increasing one variable known as a parameter. By decreasing another parameter, the cloud literally evaporates.

The above cloud equation is just one of many meteorological equations. Incredibly, if you take the complete set of mathematical equations that describes the atmosphere, you can generate a forecast. Here's how that works.

On a computer you take a full set of meteorological equations and have them describe the weather several hours ago. To do this, you feed into these equations real weather data taken from worldwide observations, also from several hours ago. Next you take the equations and you have them describe the weather now. To do this, you enter into the equations current weather observations. Finally, by having equations that are loaded with the data from several hours ago and also having them loaded with data from right now, you get a forward momentum in the math that leads you to what the weather will be in the future. This is how computer forecasts are made.

Now, as a sidebar, if you're wondering why forecasts are not always accurate, it's not because I'm in error when I say that weather is a firmly established system. It is because of many other factors. For one, data is not received from every inch of the world. As a result, when what little data we have is entered into the equations, assumptions have to be made for the data-sparse areas in between. Assumptions lead to errors. Second, our computers are only so big and so fast. In order to generate a forecast in a reasonable amount of time, not every parameter can be dealt with. As a result, some weather factors are left out of the computer programs in order for the computer to speed up its time for making a forecast. Again errors are made.

Don't let this take away from the fact that the weather is a system here on earth that behaves in a certain and describable way. There is a set amount of energy flowing into the system from the sun, and the *weather equals the way that this energy gets reshaped and dispersed.* Look, I made an equation. And that equation is the essence of system weather.

BIBLICAL SOURCE OF LAW 2

My pocket dictionary defines system as "a world, universe, or set of facts, arranged in a logical order to show an established plan." If you want an example of the ultimate system, then look no further than Genesis. That is where God's plan unfolds before us as He logically goes about setting up the entire universe. Talk about a system!

The design of system weather is embedded in the construction of this grand system for our universe. Our weather is a direct result of interaction between the sun, atmosphere (or sky), water, and land. There are other points of finesse that God threw in that really make our weather what it is, but let's concentrate on the four most necessary ingredients for weather, which are sun, sky, water, and land.

We already saw in the previous chapter that the sun was created in Genesis 1:16, "God made two great lights-the greater light to govern the day...." That is one-fourth of the components needed for making weather.

The second ingredient for making weather is the atmosphere, or sky. It is in Genesis 1:6-8, that God's developing system of creation adds in the sky, in order to interrupt the water that is initially everywhere. In fact, He uses the expanse of sky to separate the water on the earth from the clouds of water droplets above. "And God said, 'Let there be an expanse between the waters to separate water from water.' So God made the expanse and separated the water under the expanse from the water above it. And it was so. God called the expanse 'sky.'" We're now halfway to making weather.

Next, let's deal with water. In Genesis 1:2, water was everywhere, as "...the Spirit of God was hovering over the waters." Part of that water became focused into oceans in Genesis 1:9 through 1:10: "...and the gathered waters He called 'seas.'" Now, we're 75 percent of the way to making weather.

Finally, land was necessary for weather to happen. Just like the seas, land was also created in Genesis 1:9 through 1:10: "And God said, 'Let the water under the sky be gathered to one place, and let dry ground appear.' And it was so. God called the dry ground 'land'...." We now have our fourth ingredient.

Since the early days of creation, system weather has been logically set forth by God.

In Genesis, the sun, sky, water, and land are officially established. From this point on, the weather machine surrounding the earth has been able to function. Since the early days of creation, system weather has been logically set forth by God. Because of weather's interaction on the world, the rest of God's systems are also able to function.

BIBLICAL EXPANSION OF LAW 2

Law 2 states that weather is a system. In fact what a clever, intricate, massive and marvelous system it is. As big as it is though, there are other systems around us equally as large. Some examples would be human life, plants, animals, and food. These are also laid out for us in Genesis. As a result, even though Law 2 states that weather is a system, let's not be narrow-minded. We must expand Law 2 to state that *everything* is a system.

What's truly amazing is that all of these systems are not disjointed. They interact with each other. For instance, look at rain. It is part of system weather, but it also makes plants grow. So it is a part of the plant system. Animals and people eat plants and drink water, so rain is also part of the human life and animal systems. Setting up all of these systems is a miracle. Setting up all of these systems to successfully interconnect with each other is a miracle beyond miracles!

All of these miracles are set forth for us in the first book of the Bible, Genesis. Genesis is a massive collection of all the miraculous systems designed by God. Genesis is more than just a book though. It is an immense grand system itself. After all, the system of Genesis is a plan for the world, and universe, implemented by God. All other systems, including weather, are merely a sub-set of the Genesis system. Genesis encompasses every-thing we know in our universe, as well as things we have yet to learn or recognize. Nonetheless, it is an estab-lished plan. Like weather, Genesis fits the definition of a system. It is "a world, universe, and set of facts, ar-ranged in a logical order to show an established plan." In the following paragraphs, the proof of a logical and purposeful plan speaks for itself.

First, in Genesis 1:1, God made a starting point in time and space for His system. That is why Genesis 1:1 begins with, "In the beginning ...." As Genesis 1:1 con-tinues, God's creation takes form: "...God created the heavens and the earth." Initially, there was a lot of water as Genesis 1:2 maintains "...the Spirit of God was hover-ing over the waters." Our world now has a beginning to it, and logical progression of design is taking place.

God's plan of establishment continues to build in Genesis 1:3 as He replaced the darkness with light. "And God said, 'let there be light'...." The next step was to create a twenty-four-hour period with light and dark-ness. This is done in Genesis 1:5, where "God called the light 'day,' and the darkness He called 'night.' And there was evening and there was morning-the first day." God's purposeful plan of creation continues to evolve in a logical sequence.

As noted earlier, in Genesis 1:6-8, God made the sky to separate the clouds of water above from the water be-low. This was a forward step in creating the earth. Our world is reflecting the established plan of God. The grand system of the planet is taking shape.

The logical order of God's plan takes us next to Gen-esis 1:9 and Genesis 1:10. Here, the dry ground is made to interrupt all of the water on the earth. "And God said, 'Let the water under the sky be gathered to one place, and let dry ground appear.'" We now wind up with land and oceans. Earth is looking ever more recognizable to us.

The world is already quite a system, but it is far from complete. In Genesis 1:11, God said, "Let the land produce vegetation: seed-bearing plants and trees on the land that bear fruit with seed in it...." He has not only added plants, but also devised a way for them to multiply, via the seeds. In addition, He has created a food source of fruits and berries.

The planet keeps progressing under God's plan, as implemented in Genesis. By Genesis 1:16, God created the sun, the moon, and the stars: "God made two great lights-the greater light to govern the day and the lesser to govern the night. He also made the stars." Our world has now expanded into a universe.

All right. We now have time, land, water, sky, heaven, clouds of weather, earth, the sun, the moon, the stars, vegetation, food, and the crescendo keeps building. In Genesis 1:20, animals of the water and animals of the sky are made. "And God said, 'Let the water teem with living creatures, and let birds fly above the earth across the expanse of sky.'" Just as with plants, God saw to it that the fish and birds would multiply: "...Be fruitful and increase in number...." This was all part of God's plan.

At this point, the skies and waters are filled with life. So in Genesis 1:24, life was ordained on land: "And God said, 'Let the land produce living creatures, according to their kinds: livestock, creatures that move along the ground, and wild animals....'" Animals now flourished.

Finally, it was time for man. In Genesis 1:26 through Genesis 1:27, "...God created man in his own image...male and female he created them." And just as with plants and animals, God saw to it that man would multiply. In Genesis 1:28, we read "...Be fruitful and increase in number...." Humans became part of God's established and purposeful plan.

The system, laid out in Genesis, just keeps unfolding. God sets up a hierarchy in the world. What chaos it would be without order. He says to man in Genesis 1:27 through Genesis 1:29, "...fill the earth and subdue it. Rule over the fish...the birds...every living creature...I give you every seed-bearing plant (except one!)...and every tree that has fruit with seed in it. They will be yours for food." Again, this reflects a plan on God's part.

Allow me to utter those famous words we often hear on TV, "but wait, there's more!" The system set forth by God in Genesis has many more rules and components to it, but there's one that absolutely must be included. That component is God Himself. In Genesis 1:1, God created the earth. It didn't look like much yet, just a formless, empty, dark, watery mass. His next step, before making anything else though, was to place himself amidst the world. In Genesis 1:2, "...the Spirit of God was hovering over the waters...." I can't think of any other more reassuring and hopeful component of the Genesis plan than this one.

I tremble as I immerse myself in the full appreciation of what God has laid out before us. Fulfilling the definition of a system, the Bible records the creation of "a world, a universe, and a set of facts, arranged in a logical order to show an established plan" by God. Even David trembles in 1 Chronicles 16:30 as he exclaims, "Tremble before him, all the earth. The world is firmly established; it cannot be moved." You have to agree Genesis is an awesome system. It sets up quite a universe.

There is design and order to this universe that has been implemented by God. He is the man-with-a-plan.

There is design and order to this universe that has been implemented by God. He is the man-with-a-plan. His system screams perfection. If you have any hesitation in believing this, just look around you at the world He gave us. Don't dwell on the evildoings of people, for the evil is not of God. Instead, dwell on the way the world is put together, the way we are put together, and the way all has been provided for us if we would responsibly utilize what God has given us.

We should all be in awe of the intricacies of what God's hand has made. Just as with weather, there is a myriad of subsystems that God created, and they are intertwined and connected to each other in a perfect way. Their designs are truly of pure quality. They all work, and they all work well.

On the other hand, if we had been raised, void of the systems around us, and then asked to design them, there is no way we could have designed them in any way that could compare with God's architecture. As experience and history show, man's systems only have flaws and holes in them. God's creation of the world and universe, however, is flawless.

As an example, think about the way our bodies are put together. We humans are one of God's most miraculous systems. We're like a car with an intake for edible fuel, with an output for exhaust and waste, and with the heart acting like an engine pumping blood through a highway of veins, arteries, and capillaries. The miracle extends when you consider the way our skin heals after it's been cut, or the way a new life forms and grows in the womb. We're talking about an incredible system here, and a top-notch one at that.

Here's another example of a system. Think about the cycle of water that we all learned about in science class. The cycle begins with water evaporating upward from the oceans. This evaporated water eventually forms into clouds. As the clouds get swept over the landmasses, the moisture in the clouds condenses into water droplets, and falls as rain. The rain then fosters plant and animal life. Eventually, much of the rainwater seeps into the ground and drains into a network of rivers. Finally, the rivers flow back into the ocean. From this point, the cycle of water can begin all over again. This is another of God's great systems. Are you excited yet?

Try this one. Think about the system of air and the perfect way it is composed of exactly what is necessary for all life on earth to live. The formula for air is precise: 78% nitrogen plus 21% oxygen plus extremely small quantities of many other chemical elements. A small deviation in the precise formula of air would end us all. Despite that fact, a system is in place that maintains the correct percentages. If it's nitrogen you're concerned about, the correct nitrogen level of our air is preserved by decaying microorganisms in the earth's soil, as well as by volcanic exhaust. See, Mount St. Helens serves a purpose. On the other hand, plants interacting with sunlight maintain oxygen. Defaulting back to science class once again, you'll recall this is the process that we call photosynthesis. The entire system is so perfect.

> It is the perfection of these systems, and the way they in turn perfectly interconnect and react with each other, that can only lead to a conclusion that God is real, and God is awesome.

Only a loving God could have put in place the infinite number of flawless systems that we observe all around us. Because the number of systems around us is incalculable, there are too many to have been caused by chance.

Every system is necessary, though. If one system were missing, or defective, the "whole," would fall apart. Everything is too perfect. It is the perfection of these systems, and the way they in turn perfectly interconnect and react with each other, that can only lead to a conclusion that God is real, and God is awesome.

Additionally, His creation reveals so much about His character. One way to determine His precise nature is to search for the motivation behind His plan and design of the universe. Jeremiah 10:12 is a great passage for revealing just what God drew upon in order to create everything that is around us. "God made the earth by his power; He founded the world by his wisdom and stretched out the heavens by his understanding." His power, His wisdom, and His understanding thus propelled God.

The fact that God has always been with us since the beginning of creation, despite our actions, is also an act of love.

In the above passage, God's almighty power allowed Him to create the world that's laid out for us in Genesis. That degree of infinite power just can't be found anywhere else. Only God has that much power. Again in the above passage, His wisdom allowed the Lord to create a world that offers everything that is necessary for its inhabitants. Likewise, that degree of wisdom can't be found anywhere else. Only God is that wise. Finally, from the passage above, His understanding allowed Him to create a universe that reflects His love for us. Once more, that degree of understanding and love can't be found anywhere else, except in conjunction with God!

Some of you may be asking how to recognize this love of God. For one, His love is manifest by His creation of a beautiful world around us with beautiful systems in place.

The fact that God has always been with us since the beginning of creation, despite our actions, is also an act of love. In addition, His understanding and love for us is reflected by the fact that He gave each one of us free will. That is something we should never take for granted. This free will is absolutely part of His grand system for us. Using our free will, it is our choice to appreciate all that God has made. The choice that we make sure makes a world of difference in how we view life.

Cleverness, usefulness, solidity, perfection and beauty are all components of the grand system that God created around us. They indeed reveal so much of His loving character. Allow me to return to Romans 1:20, which says, "For since the creation of the world, God's invisible qualities-his eternal power and divine nature-have been clearly seen, being understood from what has been made, so that men are without excuse." All we need to do is to look around us and we will see who God is and what He is all about.

LAW 2: PATHWAY TO JESUS CHRIST

It's no secret that during my meteorological education I became astonished upon my discovery that weather is a complete system. Later I was awakened by my discovery that God is the architect of that very system. His blueprint for system weather is laid out for us in Genesis. I now recognize that the world that God created is actually a mammoth, interconnected system made up of millions of smaller subsystems. In fact, weather is merely one of many subsystems, which we continuously encounter here on earth.

Each of God's subsystems maintains its integrity when held up to the dictionary definition of a system. Earlier in this chapter it was mentioned that a system is defined as "a world, universe, or set of facts, arranged in a logical order to show an established plan." Whether you look at the grand system of creation, or at individual subsystems like weather, the definition above applies. Each of God's subsystems was intentionally planned and designed to operate interactively with other subsystems in order to form the world around us.

These subsystems are gifts we should be grateful for. In some cases this is very obvious. For example, the subsystem of air is a highly appreciated gift by us air-breathing humans. The subsystem of water is another example. However, God also put forth other systems that may not be as obvious. For instance, have you ever contemplated the system of free will, which certainly is a gift from God to us?

God did not have to give us free will. With His power, He could have done anything He wanted when designing us. Really, giving us free will is a high-level act of love. Unfortunately, most of us take our free will lightly and also for granted.

Using our free will, we can choose to take a lesson from what motivated God in His design of the universe. His motivation was power, wisdom, understanding, and love. Ideally, these are the same four items we should draw upon as we use our free will to make decisions in our homes, at work, or anywhere else. After all, human power that is not tempered by wisdom, understanding, and love, leads to many of the man-made problems we have created for ourselves.

Unfortunately, our free will is a gift that is not always motivated by the same four positive items that motivate God's free will. Deuteronomy 32:4 is a wonderful example for demonstrating to us how to apply free will in the right fashion. The passage states, "...his works are perfect, and all his ways are just." This reveals God's character and His attitude behind His system of creation. He uses His free will to only make works that are perfect. In addition, He does nothing unless it is just. Are we the same way?

Our free will comes into play again as we contemplate the greatest of God's systems, the system of Jesus Christ. Referring to the earlier definition of a system, Jesus Christ is indeed the culmination of a "set of facts arranged in a logical order to show an established plan." The set of facts is laid out before us in the Bible.

The facts are that in the Old Testament, the cycle of man's turning to God, and then turning against Him, was never ending. God's people were sinners with no way out of this cycle. Man appeared to be condemned. By God's grace, though, a plan was foretold that would come and end this cycle. The word of this plan is sprinkled throughout the Old Testament. This plan foretells the coming of a savior for mankind. That savior is Jesus Christ.

Nowhere is the system of Jesus Christ more apparent for us than in Isaiah 53. Here, the plan of God's Son is laid out as a prophecy. The New Testament is where that prophecy becomes fulfilled. Once again, the Bible can be thought of as a blueprint for us, foretelling God's plans for His children. Simply put, His system of Jesus Christ is revealed for us, and then it becomes ours to embrace.

As with any plan, though, transforming it from a blueprint into reality is not always a straightforward task. In the case of Christ, there are an infinite number of ways by which one can find Him and accept Him, in both the mind and heart. That is why we hear so many unique testimonies when we inquire how different people came to know Jesus.

I once paid a visit to one of my church's Bible study
groups that meets early Wednesday mornings. The group
was made up solely of men. A young, nonbelieving visi-
tor, who was in the early stages of exploration, attended
that morning. He was checking out the system of Jesus
Christ and was here to see if he could learn more. Let's
call him Mike.

Until that time, Mike had believed in nothing, and
his life was a mess. I sat one chair's distance away from
him, and had no trouble detecting how miserable he was.
Thankfully, he realized that there had to be something
better out there for him, and so he came that morning to
see if he could find a solution to his dilemma. He found a
solution all right, but I was as dismayed and astonished
at the answer, as I know he must have been.

In a nutshell, here's what happened. An older gentleman on my right, who I'll call Jake, confronted the young
seeker on my left. With a note of frustration and impa-
tience, Jake began by exclaiming to Mike that Jesus Christ
was the answer to all his problems and that's that. Jake
lectured, as his voice grew louder and angrier, that he
didn't understand why Mike couldn't become a believer
right then and there. With worried eyes, Mike listened
as Jake continued with a tirade, hollering that he didn't
know what more Mike wanted. Jake sternly told Mike
that if he had a problem understanding it, then he should
just accept Christ now and deal with the rest later. Jake
concluded his monologue by demanding that Mike quit
searching and just accept Christ that very instant!

As I mentioned earlier, there are an infinite number
of ways by which one finds Christ, but somehow I don't
believe Jake's way is the one that most folks would care
to choose. When we are searching and questioning and
considering, there are often three hurdles that many of
us must jump over before Christ can really reside in us.

The initial hurdle is our own free will. We first must choose to use our free will to explore the possibility of Christ. The second hurdle is to accept Christ in our minds. The third hurdle is to transfer that belief into our hearts. Only when both of these last two hurdles are successfully conquered can we get on the road to a growing, maturing, and lasting relationship with Jesus.

Coming to Jesus was tougher for me than it was for my wife, as she basically grew up in the church. You don't suppose in my case it could have something to do with the fact that I was born Jewish, do you? I was in my mid-20s when I began my search. The search started right after weather had brought me to an unshakeable belief in God.

It's been years since that early searching, but I can still empathize with Mike, because I know just what he went through. I also recognize that someone like Jake can't just order someone like Mike to start believing, at the snap of a finger. If you're a breathing human and you have any intelligence, you need to first appease your brain, and then your heart, before you can become a true believer. That often takes a lot of time, but it is the only way a healthy, spiritual life can ever be established.

In my case, I needed a lot of appeasing back in my younger, searching days. I used my free will to decide to explore the possibility that Jesus was real and could be my savior. As I searched, the initial resistance from my family tree was immense. I asked a few of my relatives to tell me why they didn't believe Jesus was our Messiah. They responded that it was because we were Jewish, and that Jews don't believe in Jesus. Somehow, their answers did not pacify my brain. To make matters worse, I could hear the same anger in my relatives' responses to me, as Mike heard in Jake's response. I gave up on getting a well thought-out answer from my relatives and started looking elsewhere.

My looking took me right to the Bible. I realized I was on my own with my searching, so I voraciously read the Scriptures every night in the solitude of my Lawton, Oklahoma, apartment. I began with the Old Testament, and then went right on through the New Testament. This was the first time I had ever really read the Bible. That is because I had always thought that the Bible was an intimidating piece of literature.

. . . initially I was quite intimidated as I launched my journey into the Bible. I was so sure from the start that this book was written for others to comprehend.

Of course, I was easily intimidated by all literature. In Junior High School, I was the guy who read Ernest Hemingway's *The Old Man and the Sea* and thought it was just a nice, and thankfully short, book about a man who caught a fish. When my teacher quizzed me about the symbolism involved, I couldn't respond. She accused me of never reading it. Apparently, the book's meaning had gone right over my head.

So yes, initially I was quite intimidated as I launched my journey into the Bible. I was so sure from the start that this book was written for others to comprehend. I was so certain that it had personally unattainable symbolism. My prejudgment and stereotyped thinking were blown away as I began to read. I marveled that I comprehended the message that God was conveying.

I now believe that when a newcomer takes the first steps to use his free will to sincerely try out the Bible, God steps in and honors that sincerity. As long as you have an open mind and an open heart, God will reveal to you great Biblical truths. Your mind and heart will be enlightened as you literally feast on a smorgasbord of divinely inspired thoughts and emotions.

After completing my first read-through of the Bible, I reminisced. I had no trouble seeing how the Old Testament seamlessly flowed into the new. In addition, I noted how the New Testament appeared to flawlessly complement the Old Testament. I also could recognize that a coming Messiah was advertised in the Old Testament, and that He would be God's loving solution to our sins. The New Testament then went on to show this Messiah to be Jesus Christ.

I was making progress. After all, weather had led me to God, and reading the full Bible had been enough to make me realize that *perhaps* Jesus was the Messiah. The pieces looked as if they fit together, but I still had a lot more concerns.

It was now time to take my concerns to God. Armed with my belief in Him, I felt comfortable praying to Him. I requested guidance that would either encourage allegiance to Christ or dismissal of Christ. I felt only God knew for sure whether Christ was His Son and our Messiah. I had confidence that God would lead me in the right direction. There is no doubt in my mind that He honored my sincerity.

God had already placed a Christian, named Russell, in my workplace. Russell approached me less than a week after my prayer to God, and asked if I would like to go to his church. I have never asked Russell what prompted him to invite me, but I do believe that he was being obedient to a thought planted by God. I accepted his invitation, mostly for the opportunity to observe Christians. I began attending Central Baptist Church of Lawton, Oklahoma. There, I watched to see how churchgoers practiced Christ's teachings, and I watched to see how they handled life. My comfort level among

Christians was soon on the rise and my mind attempted to process all that I was witnessing.

I again sought God in a late-night prayer. I told Him how much I wanted to believe in Christ, and how much I didn't want to make a fool of myself by taking an incorrect leap of faith. God did not answer me with a voice, but impressed upon my mind to try an experiment. (Perhaps this is how God talks to scientists: perform an experiment.) The grand experiment was for me to try out Christ's teachings for myself. I would implement them into my own personal practice. Theoretically, the results would speak for themselves. Either I would wind up making a fool of myself, or I would be on to something big. In any event, this was the path I felt God was leading me down.

So far, my search for the truth sounds so calm and methodical. I would be remiss if I didn't disclose that it was actually agonizing.

So far, my search for the truth sounds so calm and methodical. I would be remiss if I didn't disclose that it was actually agonizing. During the searching and grand experimentation, I often felt like I was undergoing an operation without anesthesia. This operation could eventually lead to my mind and heart being altered. The middle of the process was especially tortuous. The middle was a time when I had one foot planted in the safe haven of my original world. At the same time, my other foot was planted in my new Christian world. My mind felt tormented as it struggled and debated to maintain its hold on my old Jewishness, while at the same time grasping for my new religion. The whole process was an all-consuming ordeal. I went through bouts of sleeplessness, often questioning what I was doing. I was able to follow it through because of one comforting thought. I knew that at the end of this quest for truth, I would either be a firm believer in Christ, or I would dismiss Him completely. Either way, my internal wrestling would end.

Eventually, my internal wrestling did indeed come to an end. I had practiced Christ's teachings for about six months and my life had changed only for the better. Positive results were following my new ways. It was curious to me, though, how the results would often begin in a negative fashion. For a moment, I would then give up on my infant belief in Jesus. Given time, however, the result would rotate into positive territory, almost always in a different way than I could have imagined. This led to a sprouting of spiritual maturity.

Now I have the confidence to declare that Christ is my Messiah. I still continue to observe what Jesus is doing in my life and the lives of others. This in turn continues to nurture and build my faith. The system of Jesus is permanently a part of me. It is a system worth embracing.

My mind faith and my heart faith were now growing. Those two big hurdles of the mind and heart had thus been overcome. God had led me down a path of experimentation that he knew would lead me to the right answer. Now I have the confidence to declare that Christ is my Messiah. I still continue to observe what Jesus is doing in my life and the lives of others. This in turn continues to nurture and build my faith. The system of Jesus is permanently a part of me. It is a system worth embracing.

...FOUR...

Law 3: Weather Smoothes Out Excesses and Deficits

METEOROLOGY OF LAW 3

Our television antenna was mounted sturdily to our chimney, and it pulled in the New York City stations rather well. It was time for the news, and normally as a kid, I tuned out once the weather report was over and the sports segment began. I made an exception though when Warner Wolf would deliver the sports. His trademark was to start his sportscast with the phrase, "Let's go right to the videotape!" and proceed to show the best and most wild stuff he could find. I have no videotape, but this chapter still contains some of the best and most wild stuff I can find. So "Let's go right to the meteorology!"

It is by masterful design that the earth is in the shape of a sphere. As a result, the part of our planet that bulges out toward the sun, near the equator, receives a great deal more solar energy than the polar regions that literally curve away. You can demonstrate this for yourself by shining a flashlight beam at the side of a globe. Notice how the light strikes the equatorial region intensely. Also notice how the light can only glance by the polar regions. The sun affects the earth in the same way, as seen in Figure 3.

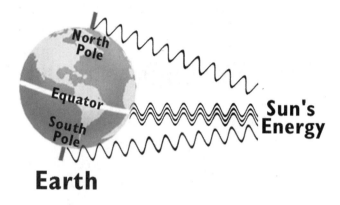

Fig. 3. Unequal heating of the earth

This ingenious design of the earth makes the tropics toasty and other locations chilly. This in turn leads to two things. Number one, it drives the great migration of chilly, retiring New Yorkers, to resettle in warm Florida. Number two, it drives the weather. Since I am not from New York, I am only qualified to expound on how the weather is driven.

It is the sun, of course, which drives the weather. Because the sun shines on a curved earth, the amount of solar energy

> Number one, it drives the great migration of chilly, retiring New Yorkers, to resettle in warm Florida. Number two, it drives the weather. Since I am not from New York, I am only qualified to expound on how the weather is driven.

striking the earth varies from location to location. Again the tropics get lots of solar energy. Heading toward the poles, that energy drops off. This translates into unequal heating around the globe.

That unequal heating is reflected by the fact that an excess of heat builds up near the tropics and an excess of cold air builds up near the poles. Opposing air masses are thus created. In meteorology, whenever there is an excess of anything in one place and a deficit of it in another, the excess moves toward the region of deficit. So the opposing air masses move toward each other, and the weather process begins. Weather is the atmosphere's way of attempting to smooth out these excesses and deficits. This is the principle behind Law 3.

For example, let's see how Law 3 operates in the Northern Hemisphere. First, the excess of cold air near the North Pole starts moving toward the regions that have a deficit of cold air. In this case, the mass of cold air would head southward. Incidentally, the leading edge of this advancing blob of cold air is simply called a cold front.

Meanwhile, the excess of warm air near the tropics simultaneously moves northward towards the colder regions. In a similar fashion, the leading edge of this advancing blob of warm air is called a warm front. Now we have excess warm air heading northward and excess cold air heading southward. Temperature excesses and deficits are in the process of being smoothed out. As a result, Law 3 explains how unequal heating from the sun gets the air masses in motion.

The story just keeps unfolding. The cold air moving down from the poles literally collides with the warm air heading north from the tropics. You basically have two air masses butting heads near ground level. The convergence of these two opposing fluids of air leads to a new river of air thrusting upward.

When air rises, the upward motion results in an evacuation of air from near the ground. With less air, the air pressure falls off. As the air pressure lowers, a system referred to as a "low" forms. This low is denoted on weather maps as an "L," which stands for low pressure.

The low is a region with a relative deficit of air. This is where Law 3 again comes into play. To fill this deficit, our original excess of both cold and warm air rush in toward the center of the low. The low literally forces a swirling and collision of the cold air and warm air masses. This is all an attempt to smooth things out and equalize the temperature.

During this process, the rising motions, and the swirling and colliding of air masses, all lead to inclement weather. Clouds form, precipitation falls, and winds blow. As a result, a low is also known as a storm center.

The opposite of a low is a "high." Highs typically bring us our decent weather. Instead of converging air masses as in lows, the ground-level air is diverging with highs. The air is literally rushing outward and away from the center of the high. Something has to replace the air as it diverges away from itself at the ground. So air gets drawn in from above. The net result is downward motions, pulling in air from aloft. This leads to a buildup, or excess of air near the ground level. So a barometer would reflect a high pressure reading, and on a weather map you would see an "H."

Putting it all together, if on a weather map there is a high near a low, you are basically seeing a region with an excess of air, near a region with a deficit of air. Law 3 states that this inequality of air will try to smooth itself out. The excess of air near the high heads toward the deficit of air in the low. We call this movement of air, "the wind," and from this the weather is born. Excess-to-deficit is a phrase that can explain most weather.

There's one more amazing piece to this weather story, though. The brilliant architecture of our planet is such that the earth spins about an axis, just like a top spins on a table. It takes an entire twenty-four-hour day for our planet to complete a full spin about its own axis. Figure 4 will help demonstrate this.

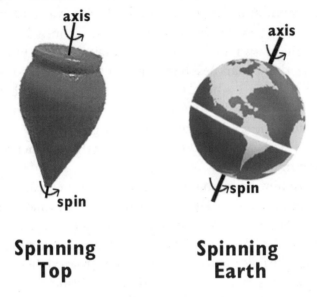

Spinning Top   **Spinning Earth**

Fig. 4. Spin comparison

A spinning earth lets the entire globe share in the sun and, thus, share in the movement of air masses. This gets the principle behind Law 3 activated worldwide. After all, if our planet did not spin, only one side of the earth would get sunshine, resulting in unequal heating, air mass movement, and weather. The other side would always face away from the sun and be uniformly cold and dark. Spinning serves to globally spread out the weather.

BIBLICAL SOURCE OF LAW 3

Weather serves as a wonderful example of the excess-to-deficit principle behind Law 3. Excesses of anything meteorological spread toward regions experiencing a deficit of that same meteorological item. This principle of excess-to-deficit is one imparted by God. It comes straight out of the Bible. The first time it appears with regards to meteorology is in Genesis.

It is in Genesis 1:14-16 that God created the sun, the moon, and the stars. Essentially, He was creating elements of excessive light. Earlier, in Genesis 1:2, we were made aware of the earth's deficit of light, by the description of darkness hanging over the earth. God utilized His Law 3, when He sent the excess of light above to the initially dark earth below: "...and let them be lights in the expanse of the sky to give light on the earth." This was a vital step in getting the weather machine in motion.

We now know that once the sun was created, other meteorological processes were triggered. First, more energy from the sun accumulated in the tropics compared to the poles. Based on Law 3, this led to hot air moving toward the poles, and cold air advancing toward the equator. Air masses were set in motion. The principle of excess-to-deficit has kept the air masses moving to this day. God's creation of the sun shining on the curved earth has always been the source of our weather.

That weather would only be confined to one side of the earth, though, unless the earth could somehow spin around. Spinning would then allow the sun to affect the entire planet. It so happens that in Genesis 1:14 we read how the "lights in the expanse of the sky...separate the day from the night." This passage reflects the fact that our earth must be spinning on its own axis. Otherwise, how else could we get day and night? After all, spinning ensures that part of our planet is always turning into the sunlight while part is always turning into the darkness. Spinning ensures that day will fade into night and night will brighten into day. Spinning may give you a headache, but it also gives us weather over the entire earth.

Once more, without the earth's spin, weather would be confined to just that side of the planet facing the sun. There would be an excess of weather on one side, and the other side would experience a deficit of weather. Thankfully God instilled a planetary spin with weather spreading all over. Excess-to-deficit is again applied during creation, and the net effect is that the weather machine is now in perpetual motion.

BIBLICAL EXPANSION OF LAW 3

God applied the principle of excess-to-deficit to more than just weather. During creation, many other systems were designed with Law 3's principle at work. As you might expect, these are revealed to us in Genesis.

One example of God's directing something that is plentiful in one region and sending it to an area where it is lacking, can be found in Genesis 1:28. After creating Adam and Eve, He told them to "...increase in number; fill the earth...." Where Adam and Eve initially stood, they were in excess. The rest of the planet was experiencing a deficit of people. So the command was given to multiply and spread out.

This principle of spreading out the excesses happens over and over in Genesis. Genesis 1:22 repeats the principle with God commanding the fish and birds to increase in numbers and to fill the empty seas and sky, respectively. Excess-to-deficit is the direction that Law 3 takes.

This may come as a shock to my seven-year-old who thinks I'm ancient, but I wasn't around during that wonderful week when God created the heavens and earth. I have my suspicions, though, that God did not make overwhelming numbers of anything. Because He commands people and animals to multiply and spread out, I believe he created just a small number of each species in a few select locations. As the numbers built up, the excess spread out.

If this is the case, then in Genesis 1:11, when God said, "Let the land produce vegetation...," He most likely made only a small number of each plant type. He then could let the plants' seed-bearing ability help to spread them out. After all, as the plants dropped seeds, new plants would shoot up in a new location around them. The fact that so many seeds can be airborne by both wind and birds also aided the spreading out of the vegetation. The excess of plants in one location would spread out toward the areas that had a deficit.

As long as we're on the subject of plants, once the vegetation spread out and grew, there was an abundance of food to be had. After all, many of the plants put forth nuts, berries, and other fruit. In addition, many of the plants were entirely edible. The plants represented an excess of food. This abundance was necessary to offset the hunger that man and animals would otherwise have. In Genesis 1:29-30, it is written, "They will be yours for food." The excess of food was ordained to be for those who would need food. One more time we see Law 3 in action.

My narrow thinking early on was that Law 3 applied to just the weather. Then Genesis showed me how Law 3 applies to people, plants, animals, and food, as well. With more consideration, I believe Law 3 can be expanded even further. God has imparted the principle of Law 3 to cover many processes we experience around us.

In fact, we witness the principle behind Law 3 all the time. For instance, if you fill a balloon with air, you have an excess of air that's pressured to stay within the confines of the expanded balloon. If you then put a pinprick in the balloon, the high-pressure air inside will quickly travel out the hole, toward the low-pressure air outside. Excess-to-deficit is in action.

A second example is a dam. You have an excess of water behind the dam and a relative deficit of water in front of it. If you put a hole in the dam, the excess of water will gush through toward the outside. Law 3 is upheld.

Here's one final example. Your car battery is filled with an excess of negatively charged particles called electrons. These electrons are just waiting to travel. When you turn the car on, completing a circuit, the excess of electrons follows a path toward areas that are electron starved. Basically, the electrons flow out of the negative side of the battery toward the positive side, where there is a deficit of negative charge. Thus the principle behind Law 3 is maintained. Excess-to-deficit is all around us.

LAW 3: PATHWAY TO JESUS CHRIST

The excess-to-deficit principle was born during God's formation of the earth. He imposed Law 3 on the setup of weather, and He imposed it when designing all of His other creations, as revealed in Genesis. Law 3, though, does not end with Genesis. In fact it is an integral part of the story line that weaves its way through the entire Bible.

Consider the fact that the Old Testament is loaded with account after account of man's wicked ways. God's children are basically loaded with sin. There is an absolute excess of sin in the people of the earth. That sin was standing in the way of a pure relationship with our Heavenly Father.

Early on in the Bible, God had His people atone for their sin, usually by sacrificing an unblemished animal. In theory, the sins would symbolically be transferred into the sacrificed animal. Man had the excess of sin and the animal had the deficit. So a net transfer took place. Unfortunately, this process did not permanently change God's children. They typically went right back to sinning.

A permanent solution was needed, so God sent us the ultimate sacrifice. He sent us His Son, Jesus Christ, to be the unblemished, sinless sacrifice. We are loaded with sin, but Jesus died on a cross and took away our sins. He did it for the people back in the Biblical days and He did it for us today. This is by far the finest example of the excess-to-deficit principle. Our excess sins transferred to the sinless one. As Christians, we can now have an unhindered relationship with God our Father, thanks to the sacrifice made at Calvary.

> **We are loaded with sin, but Jesus died on a cross and took away our sins. He did it for the people back in the Biblical days and He did it for us today. This is by far the finest example of the excess-to-deficit principle. Our excess sins transferred to the sinless one.**

Jesus Christ Himself was an espouser of Law 3. The Son of God spoke of the excess-to-deficit principle often, commanding all of us to apply it to our actions. One example of this is noted in Matthew 5:44, "Love your enemies...." We are told to spread our love to those who are basically loveless. I imagine that for the nonbeliever this would be a tough command to follow. Christians are filled with the love of Christ, however. This gives Christians the resource of excess love within, and thus enables them to give it to those who have no love.

Matthew 6:2 is another example of the principle behind Law 3 in action. The verse reads "...give to the
needy...." This is Jesus' command to spread the wealth,
from the rich to the poor. Again, I would think nonbelievers might have a tough time with this command.
Actually, many believers have a tough time with this.
We must remember that this verse applies to finances
and a whole lot more. People can be in need of more
than just money. Once more, as Christians we are empowered with the variety of riches God has bestowed
upon us. However, these riches are not meant to be
hoarded. The command is given to spread this excess
wealth to those who have not.

The concept behind Matthew 6:2 is reinforced again,
this time in Matthew 25:35-46. Here, Jesus spoke to the
righteous, saying, "For I was hungry and you gave me
something to eat, I was thirsty and you gave me something to drink, I was a stranger and you invited me in, I
needed clothes and you clothed me, I was sick and you
looked after me, I was in prison and you came to visit
me." The righteous were puzzled, wondering when it was
they had done all this for Jesus. He replied, "...whatever
you did for one of the least of these brothers of mine, you
did for me." To the unrighteous, Jesus also had this to
say: "...whatever you did not do for one of the least of
these, you did not do for me." The Son of God concluded
His speech with consequences for all behavior. He stated,
"Then they [the unrighteous] will go away to eternal punishment, but the righteous to eternal life."

These verses of Jesus encourage us to be righteous.
We are to obey the command to give to those who are in
need of food, or drink, or friendship, or clothing, or medical attention, or compassion or forgiveness. We are so
blessed with the excesses in our lives, because Jesus first
gave His all to us. These verses are a command to take

what we have in abundance and to spread it out over those in need.

The Great Commission is one other example of the excess-to-deficit principle. In Matthew 28:19, Christ tells his disciples to "Therefore go and make disciples of all nations...." The disciples, and those of us who have followed as Christians, are to spread the word. We have the excess of good news, and most of this world has the deficit. Present-day missionaries are merely following through on this principle of Law 3. They plant seeds of faith as they share with others the difference Christ can make in people's lives.

Christian recording artist, Michael W. Smith, sings the song, "Everybody's Got Seed to Sow." He is so right. All of us, believers and nonbelievers, have an infinite wealth of seed in us. We are all like the rotary spreader I use on my lawn to put down grass seed. Depending on the sower, though, not all of the seed is necessarily good. The abusive parent spreads seeds of pain to his children. These seeds take hold and grow. Those that gossip spread seeds of rumors that can tear another's reputation apart. Racists spread seeds of prejudice and hatred. Unfortunately, *everyone* has seeds to sow.

The miracle of Christ is that, as we choose to receive the seeds that He has sown, our nature changes. He has offered us a grand blueprint for living our lives. This in turn enables us to start spreading and planting seeds of improved quality. These seeds bear worthy fruit, often in roundabout ways that one can only marvel over.

As we learn to walk in the light of Jesus, and as our seeds take hold, leading to positive results, our faith in Him grows and matures. This in turn encourages us, and so it becomes very natural for us to follow His commands to spread out the excesses that we possess. With the help of Christ, we can all be doers of Law 3.

FIVE

. . . .

Law 4: An Attitude of 'Whatever it Takes' is Behind the Design of Weather

METEOROLOGY OF LAW 4

A quarter for five balls: that's what it used to cost for a game of pinball just off campus from Penn State. It was late one evening and neither my roommate, Roger, nor I, had a date. We decided we'd go downtown and parlay our eight quarters between us into hours of action at the pinball arcade. Thirty minutes later, we were quarterless and finished.

My story is that I just wasn't any good at the game. Pretty simple story, huh? Roger's story is that he'd get emotionally swept into the game. In fact, he would wind up slamming and pounding the pinball table to keep the ball alive. He believed he was giving the ball some extra action. The only action he got, though, was a display light that would turn on and illuminate the word, "tilt." His pounding and slamming had tilted the table, the game would freeze up, and the pinball would drop out of play. He thought he was doing *whatever it took* to keep the ball alive, but as you can imagine, the games went pretty fast. His response was that he felt tilting should be allowed.

Well guess what, tilting is allowed, and it's going on earthwide. The axis of our planet is tilted at an angle, like the Leaning Tower of Pisa. Furthermore, you'll recall from chapter 4 that the earth is spinning like a top. The result is that we now have a spinning, tilted planet. This makes for a cockeyed-looking earth, as seen in Figure 5.

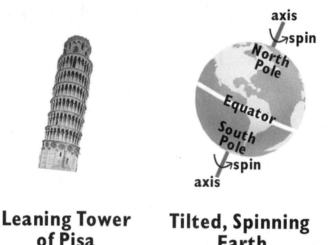

Leaning Tower of Pisa   **Tilted, Spinning Earth**

Fig. 5. Tilt comparison

Spinning and tilting are only part of the story. There is actually one additional dimension of movement that involves our planet. That movement is our earth's rotation around the sun. In fact, every 365 days, our planet completes one full rotation around the sun. So now, we not only have a tilting and spinning globe, but a rotating one as well. This entire combination of tilting, spinning, and rotating is precisely "whatever it takes" to give earth its seasons. A couple of diagrams will help to demonstrate this.

First, imagine that you are sitting in your easy chair in outer space and gazing at the earth and sun. Figure 6 portrays this scenario. Notice how the top half of the earth is tilted toward the sun. The sun's energy is focused most intently on the earth's Northern Hemisphere. As a result, the northlands are warming up. In fact, this is the precise configuration for summer north of the equator. Meanwhile, the Southern Hemisphere is tilted away from the sun. As a result, it is winter down there.

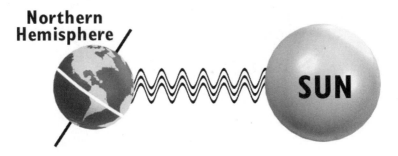

Fig. 6. Northern Hemisphere summer

As long as you're sitting in your chair in outer space, why not spend the time to watch the earth over a twenty-four-hour period. You will see it make one complete spin about its own axis. That means that as you eye the earth, and focus on a specific location such as your hometown, you will see it spin in and out of sunlight once every day.

All right, enough sitting in one spot. It's now time to take the spinning, tilted earth on a fairly circular journey around the sun. This journey is often referred to as a rotation, or an orbit or revolution. Whatever you call it, the journey takes a full year to complete. It then stands to reason that it takes half a year to travel only halfway around the sun. Let's go ahead and halt the earth's journey around the sun at the six-month point, and once again lean back in our easy chair. Figure 7 represents what you would see.

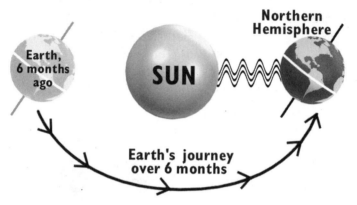

Fig. 7. Northern Hemisphere winter

In the above diagram, six months have passed since the Northern Hemisphere was aimed at the sun. Now the Northern Hemisphere is tilted away from it. It is thus winter in the northern half of the world. Meanwhile, the Southern Hemisphere is now aimed at the sun, so it is summer down there. The seasons have exchanged themselves from one hemisphere to the other, over the course of half a year.

If we sit in our chair and let yet another six months go by, the earth will have traveled completely around the sun, right back to its starting point. Once again, summer will have returned to the Northern Hemisphere and winter to the Southern Hemisphere. The complete trip to get us back to our starting point has indeed taken a full year. In the north, we have gone from summer, to winter, and then back to summer.

Are you at all wondering about the other two seasons of autumn and spring?

At first glance, our globe probably appears cockeyed and chaotic looking. After all, it's tilting, and it's spinning, and it's rotating all at one time. Looks can be deceiving, though. There is absolute ingenuity to this whole setup.

Well, three months into our initial six-month trip from summer into winter, an intermediate point is always reached. That point is called autumn. Likewise, halfway into our six-month trip from winter to summer, another intermediate point is reached. This point is called spring. Now, we have accounted for all four seasons.

Again, four seasons result from the tilting earth, orbiting around the sun. The complete revolution occurs over a 365-day period. There is absolutely no other way that the earth could experience its run of four seasons per year. Our tilted axis, our spinning earth, and our sun-centered rotating planet, are absolutely "whatever it takes" to get four seasons on the globe.

At first glance, our globe probably appears cockeyed and chaotic looking. After all, it's tilting, and it's spinning, and it's rotating all at one time. Looks can be deceiving, though. There is absolute ingenuity to this whole setup.

We already know that because of spin, our planet winds up with day and night. That means everyone gets to share the sun. The wealth of the sun's energy and the life it sustains on earth is spread out across the entire globe. There is an inherent fairness in all this. After all, if the earth didn't spin, the sun would only favor one side of the planet. Therefore, spin represents one component of the ingenious setup of our earth.

We also know that because of tilt and rotation, our planet winds up with seasons. Tilt and rotation are two more components of the ingenious setup of our planet. Tilt and rotation mean everyone gets to share changing weather. Plants thrive with their growing and dormant seasons due to this change in weather. Again, there is an inherent fairness in this. After all, if the earth didn't tilt and rotate, it would always be getting hotter on one section of the globe, while the other sections would continuously be getting colder. The seasonal variations act like a regulator spreading the sun's energy around from one hemisphere to another, placing a cap on how hot or cold it can get in any one location. Tilt, spin, and rotation are three examples of "whatever it takes" for making most parts of the earth livable for human beings.

BIBLICAL SOURCE OF LAW 4

A spinning, tilted, and orbiting earth is no less than one great example of planetary finesse. It is an impressive set of processes associated with our round planet. Scientists have greatly labored to see if there is any other potential combination of processes that could yield us our days, our nights, and our seasons of weather. They tried doing "whatever it takes" to yield the same results as we already see around us. Their conclusion is that there is no other way. "Whatever it takes" has already been accomplished.

Scientists are correct with their conclusion. In fact, the accomplishment of spin, tilt, and rotation was taken care of long ago by God. His handiwork was designed in the days of creation. It is written of in the verses of Genesis. God has an attitude of "whatever it takes."

God has this attitude, but He certainly appears modest about it. In fact, only once in Genesis is the setup of meteorological seasons mentioned, and barely at that. We all should be so modest about our accomplishments. Genesis 1:14 sets forth the seasons, days, and years as they are marked by the always changing configuration of the sun and stars. "...Let there be lights in the expanse of the sky to separate the day from the night, and let them serve as signs to mark seasons and days and years...."

Genesis 1:14's reference to separation of day and night is merely a reflection of God's implementation of a spinning earth. After all, that is the key ingredient for obtaining day and night. This is a prime example of God doing whatever it takes to get a desired result.

The same Bible passage above also refers to seasons and years. It is yet another reflection of one of God's implementations. This is where He set the earth on a tilt and sent it revolving around the sun. This tilt and revolution in turn leads to our seasons and years. Basically, God was doing whatever it takes to form the perfect setup for our planet and its weather.

BIBLICAL EXPANSION OF LAW 4

God not only did "whatever it takes" with regard to weather, but repeatedly He applied this same attitude and action to the creation of all the other components and systems on our earth. For example, in Genesis 1:9, God did whatever it took to create the seas. That verse reads "And God said, 'Let the water under the sky be gathered to one place....'" That's simple enough; He spoke and it happened. The fact that God could create all the wonders around us, including the seas, by simply speaking is like saying he had infinity at his disposal. Only with infinity at one's disposal can "whatever it takes" truly be realized. Therefore, no one but our God has the ability to deliver on "whatever it takes."

Now that the seas were created, God did whatever it took yet again. With infinity at His disposal, God applied all His wisdom and foresight and did whatever it took to create a life form that could live in the waters. The result was fish and other sea life. A simple verse reflects this infinite creation. The verse is Genesis 1:20, "And God said, 'Let the water teem with living creatures....'"

Similarly, in Genesis 1:6-8, God did whatever it took to create the sky. We don't actually know how He created it. We just know that He commanded it to form and it did. He obviously did whatever was necessary. Only with infinity at His disposal could He have done this. He is the only one who can truly create something out of nothing. God's power is beyond comprehension.

God used that same power to fill the sky with life. Again, in Genesis 1:20, we read "...and let birds fly above the earth across the expanse of sky." Elsewhere in Genesis we read of that power being focused to create animals, plants, food, the sun, and people. He continuously did whatever it took during the days of creation.

Let's go a step further now. Each of God's creations, including the weather, the seas, fish, birds, sky, plants, earth, and sun are ingeniously flawless, individual systems. Each system is miraculously made. Each system is perfect. Each system works. What's even more miraculous is that these systems are not disjointed from each other.

All of the systems God designed come in contact with each other. In fact, each system needs the others to function properly. There are continuous interactions between them. The weather and plants are interconnected. The seas and the fish are interconnected. People and food are interconnected. The interconnections are rampant. The interfaces are seamless. The infinite number of these systems that interact with each other is what makes for a living earth. Our planet works. Can you imagine being assigned the task of designing all these systems and then ensuring that they interact properly with each other? God merely did "whatever it takes" to make our planet function. His perfection of this task inspires deep contemplation.

What's amusing is that in the blink of an eye, God did "whatever it takes" for a planned result, such as the creation of weather. We, on the other hand, spend countless years of our lives in school and on the job, studying and working with the result. In my case, I spend 40 hours per week trying to forecast the way His weather will behave. Ultimately, though, it leads us right to Him. Some of us even wind up writing books about it!

LAW 4: PATHWAY TO JESUS CHRIST

It's amazing how loosely the term "whatever it takes" is utilized. I went into a local auto repair shop to have my car fixed, and there it was, proclaimed as a logo on a banner: "Whatever it takes to keep you, the customer, satisfied." That sounded good to me. I walked in with a coupon for 10 percent off any repair, and, naively, I was already satisfied.

Upon completion of my car's repair, I was given the bill and told to pay at the counter. I stood in line until my turn came.

"Whatever it takes" is a nearly impossible motto to deliver on. Oftentimes, those who utter this phrase wind up never really following through on it.

I handed the cashier my bill and my coupon for 10 percent off. She looked at it and told me I couldn't use the
coupon, because the parts that had been needed to fix
my car weren't parts that had been in stock. She informed
me that they were parts that had to be ordered from a
parts store down the street.

I spent several moments re-reading my coupon, which
mentioned no conditions with respect to the blanket 10
percent off my repair bill. Looking around at the small
store, I realized that they probably had few if any parts in
stock, and probably always ordered parts from down the
street. I then looked up at the sign with the "whatever it
takes" slogan. Some motto, I thought to myself.

I asked to speak with the manager. He gave me the
same argument that the cashier had given. I realized the
coupon was basically a scam, and it was only after pushing the "whatever it takes" slogan in their face that they
reluctantly gave me the discount. I never returned to that
shop, as they had let me down.

"Whatever it takes" is a nearly impossible motto to
deliver on. Oftentimes, those who utter this phrase wind
up never really following through on it. There is one place,
though, where the attitude of "whatever it takes" is fulfilled over and over and over again. That place is the
Bible, as we read about God's actions.

God's actions in the Old Testament were never a let-
down to me. It's true the Israelites let me down more than
a time or two, but God never did. In fact, as I read, I can't
help but be uplifted. God can be relied on to do whatever it takes for His perfect will to be done!

For example, it was God's will for the Israelites to be
freed from bondage by the Egyptians. He brought plagues
to Egypt so the Jews would be allowed to leave. He parted
the Red Sea so the Egyptian Army could not reach them.
He provided potable water where unfit drinking water

existed. He provided food where there was nothing to eat. I would say that God was in the habit of doing whatever it takes to keep His people encouraged and on the right track. He even gave them the Ten Commandments!

Unfortunately, no matter how much God came to His people's rescue, the Israelites would go back to complaining and sinning against God. In the Old Testament, a large-scale pattern developed whereby the people of Israel went through cycle after cycle of turning away from God and then turning back to Him. First His people would grumble and complain, and then sin and rebel against the Lord. This behavior would bring punishment. Then they would cry out to God to save them.

Jesus was sent to show us the way to God. Jesus was sent to show us the ways of God. Jesus was sent to give us hope, peace, joy, and eternal life. Jesus was sent to absolutely save us!

God loved His people and surely was sickened by the cycle of their turning against Him time and again. Their sins were creating a gulf between Him and them. Out of His love for all of us, He came up with a way to break the awful cycle described above. He basically did "whatever it takes" yet one more time.

This is what it took. God sent His only Son, Jesus Christ, to live among us. Most significant for us is that Jesus was sent to be the absorber of our sins. He absorbed them, and then He died for them. God had this whole process take place so that separation between God and His people, us, would not have to be an issue ever again. Our sins, which had been a barrier between God and us, were absorbed on the cross with the crucifixion of Christ. Death on that cross was the ultimate sacrifice by our Lord. There is no more poignant example of God going to the limit on doing "whatever it takes" on our behalf!

The depth of God's love, by offering us His only Son, runs deeper than any ocean. Jesus was sent to show us the way to God. Jesus was sent to show us the ways of God. Jesus was sent to give us hope, peace, joy, and eternal life. Jesus was sent to absolutely save us!

It's interesting to ponder the way God sought to save us from our sinful nature. He did it by coming in the form of a man. Even now, many folks I chat with relay how difficult it is for them to believe in a God that they claim is invisible. God has already allayed that concern. He did come to earth as a living and visible man. That way, people could relate to Him. That way, people could be witnesses of Him. God also recognized that for us to be able to relate to Jesus, He would have to endure at least the same temptations and hardships that we go through. The Bible details the times of extreme testing that Christ went through.

Human nature is a skeptical nature, though. God decided that He would need to foretell the coming of Christ early on to make His Son more readily expected and accepted. Jesus is indeed prophesied throughout the Old Testament. Ultimately, the prophecies were fulfilled in the New Testament.

Still, as I stated, people are skeptical. I believe God recognized when sending Christ that He would need to differentiate the circumstances surrounding the life of His Son. That way people would know that Jesus truly was His Son. Perhaps that is why the prophecies about Jesus foretold of some extreme and unique circumstances surrounding His life. In that way, when Jesus fulfilled those unusual prophecies, theoretically there would be no doubt who He was. Did I mention that people are skeptical no matter how great the pile of evidence before them?

I point this out about God sending Jesus to earth to emphasize that God has done "whatever it takes" to reveal His love for us, as well as to bridge the abyss that so many people feel exists between them and our Heavenly Father. After all, God came in the flesh to walk the earth. God made it possible for us to relate to Christ by having Him endure the same trials that we go through. He ensured there were witnesses to Christ before and after His resurrection. God also saw to it that the coming of Jesus Christ would be foretold in the Old Testament, so that we wouldn't doubt Christ when He was here. Additionally, God foretold Christ's coming so that we would understand why He was here. He prophesied so many things about Christ that it would be impossible for anyone else to impersonate Christ and fulfill all that was foretold. God performed "whatever it takes" on our behalf.

"Whatever it takes" is definitely an attitude and series of actions that humans rarely perform or witness from each other. It typically implies going to quite an extreme to get a result. God, however, has no trouble upholding a "whatever it takes" attitude and going to an extreme. He demonstrated this when He went on a mission to save His people by sending His Son to be our savior. "Whatever it takes" is an extreme action, so on a scale from one to ten, everything must be a ten. Jesus' birth was certainly a ten. Jesus' life was a ten. Finally, Jesus' death and resurrection were both tens as well.

As wonderful as this is, it's only half the story. It is true that God has done "whatever it takes" to save us. However, we also have a share of the responsibility. With the gift of free will that God has given us, we can choose to embrace God's plan of salvation for our lives, or we can choose to turn against it. Yes, God is the great example of doing "whatever it takes," but we still have to do "whatever it takes" as well.

....SIX....

Law 5: Balance is a Characteristic of Weather

METEOROLOGY OF LAW 5

Friday night had finally arrived in the middle of a Penn State trimester. The Nittany Lion lady gymnasts were competing over in Rec Hall and were ranked as some of the best in the nation. A couple of us went over to see what a competition was like. I ain't no hick from the backwoods of Jersey (yes, New Jersey does have some woods), but I felt like I was at the Olympics. These women were high-caliber athletes with impressive abilities. Most of the time their routines were nearly perfect and the judges would give them a 9.7 or some such score.

I always thought, though, that the gymnast who had a moment of floundering and then recovered, actually deserved more points. I say this because for the gymnast to recover, some pretty quick thinking and fast reaction had to occur. That to me indicated even more talent than what was exhibited in a flawless routine.

Here's an example of what I noticed. If the gymnast was on a balance beam and started to tip to the right, she would lift her left arm or left leg to balance herself out. If she came out of a tumbling routine and landed on her feet leaning forward, she would quickly arch back her body and her head and regain her balance.

The atmosphere is like a gymnast. It too strives to maintain balance. For example, if air is rising in one

The atmosphere is like a gymnast. location, you are guaranteed that somewhere else in the world, there will be an area of sinking motion to compensate for it. In this way, the weather keeps itself in balance. The net result is that there is an equilibrium that is maintained globally.

If this equilibrium were not maintained, then the atmosphere would be totally chaotic, unpredictable, and tend toward unimaginable extremes that would be unlivable. For instance, what if our atmosphere had only upward motions, and no compensating downward motions? The world would then be totally covered by low pressure. The atmosphere would turn into a global storm, getting more severe with every passing hour, day, month, and year.

However, if sinking motions ruled the atmosphere, with no compensating upward motions, we would have global high pressure. High pressure tends to be sunny and dry, so it would never rain again. The world would get drier and drier over time, turning the earth into a desert. Balance in the weather is necessary to maintain the earth we all know and love.

The rising and sinking motions described above make up just one example of numerous couplets-of-opposites that exist in meteorology. These couplets can be divided into two groups. The first group contains couplets whose components are occurring at the same time. For example, the couplet of rising and sinking motions falls into this category. If you can find rising motions on the planet, then at the same time there is guaranteed to be a region of sinking motions as well. The two motions are thus occurring concurrently.

There are the couplets of cool air and warm air, dry air and humid air, cold fronts and warm fronts, low pressure and high pressure, updrafts and downdrafts, floods and droughts, cold waves and heat waves, cloud cover and clear skies, north winds and south winds, east winds and west winds, and so on. Again, at any given moment, when one component of any of these couplets exists, then its opposite component is also occurring.

There is a second group of couplets-of-opposites made up of components that follow each other in time. Balance is thus reached after time has passed. Winter and summer in the Northern Hemisphere provide one example. The two seasons balance each other out, but it takes the course of a year for the two seasons to have occurred. So, over time, a balance has been reached. Other examples of couplets-of-opposites that balance each other out over time would include El Nino and La Nina, spring and fall, day and night, and so on.

Whether the components of a meteorological couplet-of-opposites occur simultaneously or over time, the net result is always the same. Each component is basically keeping its respective opposite component in check. Each component is literally a countering mechanism to its opposite. A balance in the weather is thus reached and maintained on a planetary scale.

BIBLICAL SOURCE OF LAW 5

Balance is observed in the atmosphere, but more importantly, it is a principle of weather that was set forth in the time of creation. In the first verses of Genesis, we literally see light shed on the true origin of Law 5. In Genesis 1:2 it is revealed that initially on earth "...darkness was over the surface of the deep...." One verse later, God balances out this darkness by giving the earth light, "...Let there be light...." Thus, the first balanced couplet was created. God even labeled the components of this couplet as day and night.

The creation of the sun as our light source comes to the forefront on the fourth day of creation. In Genesis 1:16 we note that "God made two great lights—the greater light to govern the day...." You'll recall that all of our weather is triggered by the sun. Therefore, creation of the sun in turn leads to creation of weather and all of its associated couplets. In one decisive step, God has created all of the meteorological couplets listed earlier in this chapter. These pairs of opposites encircle the globe. They reach equilibrium with each other and a worldwide balance is established.

BIBLICAL EXPANSION OF LAW 5

God's system of balance is not just confined to weather-related items. In fact, balance is an identifying trademark of most of God's work. It is His fingerprint, if you will. Throughout Genesis and the rest of the Bible, the balance theme prevails. One way to detect His signature of balance is to search out pairs of opposites in the Scriptures.

The Almighty's balancing act begins right away in the early chapters of the Bible. In Genesis 1:9, we read

"Let the water under the sky be gathered to one place, and let dry ground appear." In this one command, we read of God's creation of a new pair of opposites. We now have landmass surrounded by a great expanse of water. God followed this by evenhandedly putting life on the land as well as in the water.

On the sixth day, God created quite a pair of opposites. This is when He created man and woman. I can tell you that after ten years of marriage, I am thankful to have a woman balance me out. Lord help me if I didn't!

Marriage is definitely a lot of work, but so was creating the heavens and the earth. On the seventh day, God chose to create yet another pair of opposites, work and rest. Genesis 2:2 states "By the seventh day God had finished the work he had been doing; so on the seventh day he rested from all his work." Work balanced by rest is hopefully a balance that we all have in our lives. After all, it was initiated and practiced by God Himself.

Balance quickly became a normal component of life on earth. In the Garden of Eden, the purity, innocence, and sanctity of everything up to that point became balanced out by cunning and seductive evil. The couplet of good versus bad was born and thrives to this day.

Later in Genesis we note that pairs of opposites spring up all over. Cain and Abel are one example. Joseph and his brothers are another. We move into Exodus, which is an entire book devoted to two opposite groups. There are the free, but idol-worshipping Egyptians, and the enslaved, but God-believing Israelites.

The examples of opposites that balance each other are everywhere in the Scriptures. They are in the Old Testament and they are in the New Testament. The one pair of opposites that most stands out in the reading of the Bible, though, is the couplet of good and evil. Bible story after Bible story focuses on a theme centered on good versus bad. This couplet of struggling and battling opposites is a thread that winds through both the Old and New Testaments and continues right into the current moment.

LAW 5: PATHWAY TO JESUS CHRIST

It is in this current moment, yes, right now, that all of us are wrestling with choices we must make. There are couplets-of-opposites before us in every direction. We are faced with the often arduous task of gravitating toward one side or the other of each of these couplets. We are deluged with making choices from the moment we wake up to the moment we go to sleep. I suspect we are even making choices while we sleep, as reflected by the dreams of our slumber. Making choices consumes our lives.

Some choices we make cause little impact on the world around us. For example, everyday, most of us choose what clothes we will wear. Will it be the blue outfit or the gray one? The consequences are nil, but nonetheless, we mull over the choices and decide on a preference.

Other choices we make are based on personal tendencies. For instance, some of us tend to be hot weather people, others are cold weather people. Each group balances the other out. Except for the occasional battle at the thermostat, barely a stir is made in the world by the preferences between these opposite types of people.

We also tend to polarize our preferences, thus affecting choices we make. As an example, most people have a polarization toward either day or night. In fact, most of us classify ourselves as either a day person or a night person. This reflects the time of day or night in which we believe we function best. Balance is reached by these two opposite groups of people. Keep in mind, though, that the consequences of the polarization of either side of this couplet barely put a ripple in the pond.

However, the fact that we are ceaselessly making choices implies two highly important things about us and the world in which we live. One is that the gift of free will that God gave us is constantly coming into play during our decision-making. Another is that because God set forth balance as a characteristic of all of His creations, not just weather, choices will forever be facing mankind. After all, balance means pairs of opposites exist, and pairs of opposites means choices must be made as to which side of the proverbial coin to choose.

Certainly it is not earth shattering when we choose to gravitate toward being hot weather people versus cold weather people, or we sway toward being night people versus day people. There are other choices to be made, however, that have far greater consequences.

Let me raise the bar a notch here, and in the following paragraphs offer some examples of daily couplets-of-opposites we encounter that are more substantial. These couplets force us to choose one side or the other. The choice we make, as the road before us forks, is more consequential than deciding what color clothes to wear.

Consider what might be an average day for us. The alarm clock blares, awakening us at 5:45 A.M. We immediately encounter our first pair of opposites to choose from. Do we take on the attitude of "Thank you God for another wonderful day," or do we take on the attitude of "That alarm clock needs to be flung against the wall. Watch out world because we're angry"

As the day unfolds, we encounter more pairs of opposites, and so more choices have to be made. We're headed to work and a car wants to merge into our lane. We can choose to be nice and let them in, or choose not to be nice. Yes, to be or not to be, that is the question. Once at work, we can choose to give 100 percent to our work, or work with less than our all. We then take a break and go out for lunch. We can either order a healthy salad, or we can choose the three-quarter-pound, triple stack, bacon cheeseburger deluxe, with large fries and a thirty-two ounce milkshake. Now we head back to work, where someone tells an off-color joke. We can either laugh along with them, or stand out by not finding it funny. On the way home from work we stop by the cleaners to pick up our cleaning. At the cash register, we are given too much change. We can either keep quiet, or we can choose to be honest and give back what is rightfully theirs. Next, we arrive at home and encounter our spouses. We can either be up for them, or down for them. We can also choose to be kind to them or to be cruel to them. In less than twelve hours, the daily cycle starts all over again.

As you can imagine, there are a myriad of daily choices to make. The choices, though, are usually fairly clear. We are typically faced with choosing from a pair of opposites. I can either do this, or I can do that. Again, God put balance in this world. Balance is composed of pairs of opposites. Pairs of opposites imply choices to be made. The previous paragraph's examples of choices to be made are much more important than whether you choose to be a day person or a night person.

The couplets-of-opposites that we are subject to come racing toward us like an avalanche. The question is to which side of each couplet do we gravitate? Which side do we choose? God the Father is the ultimate role model for showing us, His children, what choices to make. Thankfully, the example that God sets for us is recorded throughout the Bible. Scripture is the perfect guide for us as we confront choices throughout our daily walk.

In the next chapter, I cite several passages from Genesis where God has just finished creating items in this world. I then point out that in the Scripture there tends to be a concluding phrase to each creation. That phrase is "...and He saw that it was good." For God, choosing good over bad really isn't even a choice, for His nature is only good. Thank goodness God's nature is solely good. After all, God ultimately has the power to do whatever He wants. Thank goodness during creation He chose to see to it that everything was good. Thank goodness that in our present day, God continues to choose that all that is of Him is good. Psalm 118:1 even encourages us to "Give thanks to the Lord, for he is good...."

"He saw that it was good" reflects God's attitude toward all that He chooses to do. God wants us to have that same attitude as we make choices in our lives. Our Heavenly Father is our role model, and this is the way that He is encouraging us to follow. We are to choose good. The tendency to do this can only come from a well-maintained relationship with God, which includes studying His word in the Bible. After all, every topic from how to be a good husband or wife to how to treat the poor, and from how to run a business to how to deal with the world, can be found in the Bible. That is why I love to refer to the Bible as a blueprint for life. It leads us to making better choices, choices that are good!

Throughout the Scriptures, there are countless verses describing consequences for choosing to follow or not to follow God's path. Oftentimes, we are like the child who has been told that it is bad to touch the hot stove, but we do it anyway. The Bible spells out the consequences either way, and incredibly, they match up to what our experiences are. As stated earlier, we tend to gravitate toward one side or the other of a couplet. The word of God, found in the Bible, is the ultimate guide and advisor. With His

Oftentimes, we are like the child who has been told that it is bad to touch the hot stove, but we do it anyway. The Bible spells out the consequences either way, and incredibly, they match up to what our experiences are. word, we are equipped to choose the winning side of the greatest pair of opposites of them all, the couplet of good and bad.

Good and bad are entities that are constantly fighting with each other in our fallen world. Together they form a couplet-of-opposites. Together they are in competition with each other.

God is the leader of good. Satan is the leader of bad. As Satan plants his seeds of evil in the minds of many people, the force of evil grows great. Without God, no matter how good we want to be, whether collectively or on our own, we will continue to fall short. Our sinful nature will always allow the evil in this world to outweigh the good. On our own, and as a group, we cannot outflank the satanic forces that naturally prevail. Only with the help of God can the world's evil be balanced by any good.

Our planet is a battleground for spiritual warfare. Sometimes Satan appears to win in worldly terms, even with God on our side. However, God has given those who believe an extra tool to ensure that an ultimate victory over Satan does occur. That tool is Jesus Christ, the Son of God. Our ultimate victory does not occur on the playing field of earth. It does occur, though, at the moment when Satan's sting would otherwise be greatest. I am referring to that moment when each of us stops breathing. Death, and the world that Satan would lead us down to, are no longer an option for the believer in Christ. God directs our path upward, as He rewards our faith. That path is an eternal one that carries us into the kingdom of heaven, next to the throne of God.

Yes, God put balance in this world. That implies couplets, or pairs of opposites. That in turn implies options from which to choose. Simply stated, the most pervasive couplet we encounter daily is the couplet of good and bad. We can attempt to keep our intentions pure and try to choose good and to be sinless. However,

> **Our ultimate victory does not occur on the playing field of earth. It does occur, though, at the moment when Satan's sting would otherwise be greatest. I am referring to that moment when each of us stops breathing.**

even if we are believers of God, our quest to be sinless will always fall short. Our free will and inherent sinful nature invariably lead us to our downfall. The Old Testament attests to this point very well, with the countless cycles of God's children falling short. That's right, even with knowledge of God, God's children ultimately falter.

At first glance, it would appear we are in a losing situation. Even with God, we fall short. God is loving, though, and God is good. He has remedied our losing situation by giving us the gift of His Son. With Christ, we no longer fall short. The greatest choice in our lives that we then must make is whether to follow God's ways and accept Christ, or to follow the world. For each of us, a personal choice must be made. There is no other choice that we will ever make where the stakes will be higher and the consequences greater.

SEVEN

●●●●●●●●●●●

Law 6: All Weather
Serves a Purpose

METEOROLOGY OF LAW 6

Moving day had finally arrived. My wife, son, U-Haul, and I would finally be leaving our small rental home for the wide expanse of our very own, brand new, three-bedroom house on one-seventeenth of an acre. The move was just eighteen miles, but it was enough distance to take us out of the local phone company's area of jurisdiction. On our way out of town, we stopped by the telephone company's main office to pay our final bill and to collect the security deposit we had made so long ago. At least, those were our intentions.

The teller was pleased enough to collect on our final bill, but she told us that we'd have to wait half-a-year before our deposit was returned. Of course I asked why, and of course she responded the way anyone would respond after being asked the same question by probably every customer who had ever left her district. In a flat, monotone, tired, and unemotional voice that could easily have been a recording coming from a machine, she said, "Because that's the way we do it. We've always done it that way." I didn't have time to argue about it, because, don't forget, we had "an adventure in moving" ahead of us and that took precedence.

One week later, the postman actually put something in our new mailbox, and the three of us tripped over ourselves racing to the curb. The world was in contact with us after the move and that was a great feeling. I opened the mailbox, and there was a correspondence from our old phone company. It was a bill for zero dollars, and it said payment was due within three weeks or we'd have to pay an interest penalty. We shrugged, ripped it up, and threw it in our recycling can.

Several weeks later, the mailbox again beckoned, and there was a new correspondence from our old phone company. The note said that they had no record of our ever paying the zero-dollar bill from a few weeks ago, so a penalty of zero was being added to the bill, and would continue to accumulate, until they received payment from us.

Using the service of our new phone company, I dialed our old company to see why they were sending so much mail about a zero-dollar phone bill. It was a different woman than I had spoken with before, but her message gave me a feeling of deja vu. "Because that's the way we do it," she said. "Until your account is closed out, and that won't happen until we send you your security deposit, we will keep sending you bills. Sir, it would help if you just mailed back the bill stub each time you get one over the next six months, and that way we would make note that you paid the zero-dollar bills. Then we wouldn't have to send you late notices, and then we would be able to send you back your security deposit after 6 months, rather than delay it." I asked her if she heard herself, and she mirrored that with, "Yes, now do you hear me?"

So for each of the next six months I dutifully wasted thirty-two cents on postage to mail back bill stubs with payment enclosed of zero. Like clockwork, the half-year passed and we received our security deposit back, and the bills of zero ceased coming. It was a senseless six month back-and-forth correspondence, but that's the way it apparently "has always been done." To me, it served absolutely no purpose.

If it's purpose you want, look no further than the sky. Every bit of weather serves a purpose, and that purpose is always a positive one. Interestingly, folks have a tough time believing that this is the case. In fact, when I speak to groups, this concept usually draws some of the most discussion. They can't believe that a hurricane, or a flood, or a lightning storm, or a tornado, or a snowstorm could serve any positive purpose.

I probably don't have to convince you that rain serves a purpose, as does sunshine. Obviously it'd be impossible for life to exist without rain and sun, so both of those elements are, without a doubt, positive. So let's jump to the severe weather. In fact, let's go right down the list from the preceding paragraph.

We'll start with hurricanes. These ferocious tropical systems are born over the warm waters of the ocean, usually in areas where the sea temperature is 80 degrees or warmer. Warm water is a potent form of energy, and so the warmer the water, the more energy it can contribute to a growing storm. In fact, hurricanes literally feed off the warmth of the ocean.

In reality, most of the world's waters are not at or above 80 degrees, so when there is a pocket of ocean that warm, it is in an area that contains a relative excess of energy. Once a hurricane forms over these tropical waters, it tends to move toward cooler regions, taking this excess energy and warmth and translating it toward areas that have a deficit. Basically, too much energy builds up in the tropical regions from the more intense sun, and the tropical storm is the mechanism for taking that energy to areas where there is a relative deficit. The storm is also the vehicle for dissipating that excess energy. The harder the wind blows, the more energy is dissipated. These are some of the hurricane's purposes.

In addition, tropical storms often bring much needed rainfall to areas that were experiencing a deficit. Just last week I was reading through a CBS script for some video of Hurricane Boris that aired across the country. The script described Boris' winds and rains around Acapulco, Mexico, and then concluded with the line, "Although the storm was destructive, portions of Northern Mexico welcomed the rain after suffering four years of drought." A purpose was served.

Let's talk floods. It is actually healthy for flood plains to experience floods. The excess water literally cleans out the soil. As the floodwaters roar downstream, the river gets flushed out. Impurities that had collected are swept away. Then, once the river returns to normalcy, it is healthier and more habitable for aquatic creatures. On top of all this, floodwaters deposit a new layer of rich soil on the landscape. This too is healthy, because as soon as the waters recede, the resultant soil is much more fertile than it used to be. Thus, floods do serve a useful purpose.

How about lightning storms? The average thunderstorm contains more energy than the bomb that fell on Hiroshima. There needs to be a way to release this energy, and lightning is one such way. An excess of negatively charged electrons builds up in one location of the storm, and there is a deficit of them in another. The result is a lightning strike. Lightning helps to diffuse the storm's energy. That is part of its purpose.

Lightning also creates ozone at high altitudes. Rising motions then transport this ozone into the ozone layer. This in turn partially offsets the depletion of the ozone layer, which has been triggered by humans.

Interestingly, a by-product of lightning flashes is also nitrogen. Nitrogen is not only a necessary component of the air we breathe, but it is also a component of fertilizer. Thus, every time an electrical storm does its thing, plants are being given a literal dose of "Miracle Gro®."

If a thunderstorm gets severe enough, it can spawn a tornado, which is usually attached to the rear of the storm. In areas of the world like the central United States, monstrous amounts of energy are tapped by developing thunderstorms. These storms grow and grow and grow. The wind and rain and lightning are not enough to dissipate the storm's energy. The result is tornado formation. It defies logic to call a tornado a storm's safety release valve, but that's what it is. If the thunderstorm's energy becomes too great, the tornado serves to take that excess energy and dissipate it via its great roaring winds. That is its purpose. Without the tornado, the thunderstorm's energy would continue to build, and who knows what kind of megastorm that would create!

Finally let's discuss snowstorms. The great snows that fall each winter serve very useful purposes. For one, they build up the water supplies for the dry months. As the snow melts during the warm seasons, the melting snows fill up streams and reservoirs. In Seattle, for example, 70 percent of the city's water supply comes from melting snows over the Cascade Mountains. June through August in the Pacific Northwest is extremely dry, but the melting snows maintain stream flow.

Another benefit of snow is that it protects crops in the winter. The wheat fields of Kansas are much more

Unlike many nonsensical things that we do in life, weather does serve a purpose, and one that is positive. productive when a blanket of snow protects them all winter from the harsh cold.

Unlike many nonsensical things that we do in life, weather does serve a purpose, and one that is positive. Admittedly, it is unfortunate that this positive purpose is sometimes "clouded" over by some of weather's negative consequences. For example, in the case of a flood, how many times have we seen people's lives thrown into disarray as their homes and businesses are washed away? On the other hand, how many times have those businesses and homes been built right on the flood plain?

BIBLICAL SOURCE OF LAW 6

Once again, we turn to Genesis to discover the Biblical source for one of our laws of meteorology. In Genesis 1:14-18 we find the initialization of weather with a positive purpose. These are the passages that reflect God's design for the sun, moon, day, night, and the seasons. You'll recall that these are also the building blocks of our weather. It is from the last line of these verses, however, that we derive the Biblical source of Law 6. This is where it is written: "...And God saw that it was good." You can bet that if God saw that it was good, than surely a positive purpose for our weather is implied.

I've read the Bible many times, and each time I have filed in the back of my mind a few verses that I found curious. Until recently, Genesis' repeating verse of "...And God saw that it was good" was one such verse that I puzzled over. I basically wondered what the necessity was of having that verse appear so many times in Genesis.

It is only as I write this book that the full picture is being revealed of what this statement signifies.

"...And God saw that it was good" is a very powerful statement. For one, it reveals the positive purpose of God's creations. For another, it reflects His personality, as well as His pure intent and love for us. After all, He certainly had a choice in how He designed all the building blocks of weather, along with His other creations. Only a loving God would have seen to it that it was all good. Again His beautiful character is portrayed.

There is more, however, behind the power of Genesis' repeating phrase, "...And God saw that it was good." In His infinite wisdom, God was well aware of the existence of a contrary figure. I am referring to the dark and evil spirit of Satan that haunts and plagues the earth to this day. When Moses was divinely inspired to write Genesis, he had plenty of experience living in a world with an evil side to it. As a result, the statement "...And God saw that it was good" allows God's creations to stand out in written contrast to the evil creations that Satan has been generating in the world around us.

Out of Genesis comes the creation of weather. Out of "...And God saw that it was good" comes the stamp of approval and positive purpose that only a loving God would place on His creations. It is as if "...And God saw that it was good" is a divine blessing placed on each creation, one that we should never take for granted and one that we should revere.

BIBLICAL EXPANSION OF LAW 6

Of course weather is not the only creation in Genesis that God designed with a purpose that was positive. He put His "...And God saw that it was good" stamp of approval on much more. In fact, in Genesis 1:3-4, one reads,

"And God said, 'Let there be light,' and there was light. God saw that the light was good, and he separated the light from darkness." Positive purpose has thus been injected into the design of day and night.

This process of setting up items which reveal positive purpose is repeated in Genesis 1:10. "God called the dry ground land and the gathered waters he called seas. And God saw that it was good." Now we have not only time marked by day and night, but also God's addition of land-masses and waterways.

Probing deeper into Genesis, we find more items that God designed with purpose that are good. For example, in Genesis 1:12, "The land produced vegetation: plants bearing seed according to their kinds and trees bearing fruit with seed in it according to their kinds. And God saw that it was good." At this point God has given the gift of plants and given them the ability to reproduce. There's nothing negative about any of this. It's all purposefully done with a positive air about it!

Purpose by God, along with His signature of goodness, keeps developing as we move through Genesis. In Genesis 1:21, more creation is revealed. "So God created the great creatures of the sea and every living and moving thing with which the water teems, according to their kinds, and every winged bird according to its kind. And God saw that it was good." In a nutshell, God purposefully created birds and fish and He made sure they were good.

In Genesis 1:25, animals are added into the mix. "God made the wild animals according to their kinds, the livestock according to their kinds, and all the creatures that move along the ground according to their kinds." The fact that this passage immediately concludes with "And God saw that it was good" implies a creation that can only be purposefully made, and one that is indeed good.

A climax of purposeful and positive creations is reached in Genesis 1:27-31. These are the passages where man and woman are created. We are commanded to be fruitful, to subdue the earth, to rule over land animals, birds, and fish, and to enjoy the providence of food. At this point, God takes a step back and examines the full spectrum of all His creations. Genesis 1:31 states "God saw all that he had made, and it was very good."

I hope by now the purposefulness of all of God's creations is apparent. There was nothing designed by chance. God made each creation with intention and foresight and wisdom and love. His accomplishments were all declared as good.

Even if you are a skeptic who doesn't want to take the Bible's word for it that

> **Look at yourself in a mirror and contemplate the way your skin heals, your heart beats, your brain processes, and your emotions stir. How could anything less than God have created all of this?**

these accomplishments are good, then all you need to do is look around. God's great handiwork abounds before us. Think of the way weather works. Think of the way our planet is devised. Examine a tree. Watch a bird. Look at yourself in a mirror and contemplate the way your skin heals, your heart beats, your brain processes, and your emotions stir. How could anything less than God have created all of this? Are these creations anything short of good? The answers to these questions can only lead to recognition that all creation is of God. It is all purposefully made, it is all good, and it is all a masterful accomplishment.

By its very nature, an accomplishment is a result that serves a purpose. To be a great accomplishment, it must serve a positive purpose. Great accomplishments are all that God has ever put before us. They are written of in the Bible, they are seen in every direction we look, and they can be spiritually encountered as we learn to walk with Him.

Because positive purpose is an unwavering trademark of God's work, we ought to have 100 percent confidence, trust, and peace in Him. Unfortunately, we often waver in our confidence in what God is doing in our lives. We sometimes go as far as to worry that His purpose may not come to pass.

These worries can be eradicated by Biblical verses that come to our rescue. In Isaiah 14:27, we read "For the Lord Almighty has purposed, and who can thwart him?" This says to us, God's purpose will indeed prevail. Three verses earlier, God verifies his firmness in purpose in Isaiah 14:24: "Surely as I have planned, so it will be, and as I have purposed, so it will stand." That sounds like unshakeable purpose to me.

Other verses throughout the Bible reinforce God's unwavering purpose in all His ways. One of these is Job 36:5 which states "...he is mighty, and firm in his purpose." Still other verses give us the knowledge, comfort, and encouragement to know that the characteristic of God's purposefully made creations is always positive. One such verse is 1 Timothy 4:4, which says, "For everything God created is good...."

Scripture is full of testimony to God's purposefulness. It is also full of verses affirming the positive nature of this purposefulness. These verses are then substantiated by what we observe around us. Everything God created, including ourselves, was made on purpose and was intended only for good.

LAW 6: PATHWAY TO JESUS CHRIST

It's refreshing to know that man was made on purpose and intended for good. In fact, this positive purpose that God had in mind for us is actually a trademark of all of His work. Out of His love, He purposely gave us minds, hearts, and once again, free will. The result is that we have the ability to think, feel, and make choices. We are anything but robots. God's intent for us is a pure one, but using our free will, we often make rebellious choices. This then leads to a distraction from the Lord's purpose for us.

I have discussed free will often in previous chapters, but once again the concept comes to the forefront. Our free will means we have a choice in how we conduct our lives. We have the option of fulfilling God's positive purpose for us. We also have the option to choose a path that strays from God's positive purpose for our lives. God's purposeful intent is always there, but it is up to us, using our free will, to bring His purpose to fruition in our lives.

Using the Bible as a blueprint for our lives, many scriptural examples stand out, exemplifying both the positive and negative results of our actions. Reading God's word, it becomes obvious that God is looking out for us. He wants to guide us. He wants our purpose to be the same as His purpose. In our lives, following His guidance makes the difference between success and failure. That may not always seem apparent, in the short term, as the world throws setbacks our way. Life is a long race, though, and in the long term, victory is assured. Our purpose needs to mirror His, and the reward will be ours.

One of my favorite passages for illustrating the above concept comes from Acts 5:38-39. In this portion of the Bible, Gamaliel the Pharisee is lecturing and educating the Sanhedrin who are holding the apostles as prisoners. Gamaliel's wisdom and life experience shine through when he states, "...Leave these men alone. Let them go. For if their purpose or activity is of human origin, it will fail. But if it is from God, you will not be able to stop these men; you will only find yourselves fighting against God." Is your purpose of human origin or is it from God?

Do you feel you even have a purpose? Because I work in the news business, I see and hear more than my fair share of video and audio from every corner of the world. Many times, newscast reports are about people who have fallen from human grace. They may have swindled, cheated, murdered, lied, or whatever. When a reporter catches up to them, the first question is invariably, "Why did you do it?" I cannot tell you how many times that I have heard them respond, "I don't know."

In my mind, there are usually two explanations for the response of "I don't know." One explanation is that people are aimless, having no direction in life, no self-worth, and no purpose. Perhaps they have never considered that God intentionally made them. This fact alone would elevate their worth to infinity. In addition, God indeed does have a purpose for them, only they must choose to accept that purpose.

Perhaps they have never considered that God intentionally made them.

The other equally plausible explanation for responding with "I don't know" is that they are avoiding having to tell the truth. That truth is that they knowingly followed a negative path, one that strayed from God's purpose for them, but are too ashamed to say so. Whatever the explanation, they are not fulfilling the positive purpose and potential that were destined for their lives by the Almighty. Basically, they chose not to seek or follow the purpose and potential God had prepared for them.

Thankfully, it is never too late for any of us to tap into that purpose and potential that God has provided for us. The centerpiece of this purpose is God's gift of His Son, Jesus Christ. Through Christ, we are able to lay down our sinful habits, be forgiven for our sinful ways, and drink out of the well of life that God designed for us. It is all there for us to choose.

It is always *our* choice. Christ stands ready to assist us in focusing our lives in a positive direction, but it is our choice. Once we start doing God's will, His purpose becomes our purpose, but it is our choice. Our many failures of our own making can yield to successes measured on a heavenly scale, but it is our choice. We can become empowered by the Holy Spirit to withstand all that the world throws at us, but it is our choice. Hope can rebound in our lives, but it is our choice. Our lives can take on pure and meaningful purpose, but it is our choice. I asked earlier, do you feel you even have a purpose at all? With Christ, your answer can only be yes.

EIGHT
●●●●●●●●●●●●

Law 7: Weather Contributes to Entropy on Earth

We've been on a journey. The previous six chapters have taken us on a trip through six encompassing laws of meteorology. We have also expanded these laws to see how they apply to all of God's other creations, not just weather. In addition, we have used them as a pathway to Jesus Christ. My hope is that the tenor of the previous chapters has been a positive one.

Before I conclude this book, I need to raise the issue of yet one additional law. I am referring to the law of entropy that exists here on earth. Entropy is a fancy sounding word, but basically it translates into a law that states that everything on earth is in the process of degrading and falling apart. This law is not as upbeat sounding as the previous six laws we have discussed. Nonetheless, this seventh law does exist and is worthy of discussion.

In this chapter, entropy will be discussed in great detail. We will see that weather is an example of a system that aids entropy. We will also see that weather is not the only driving force behind this additional law of decay.

Thankfully, as we plow through this law, I believe you will find that entropy does not have to be depressing. In fact, we already have the tools at our disposal to counter this law of the world.

METEOROLOGY OF LAW 7

It was the fall of 1979, and another weekend was about to start. We had all just finished a week of quizzes, difficult homework, and long study hours. All of us were ready to take a break, even if that pause would only last a Saturday and Sunday. We were grateful for even short breaks from the intensity of college work.

As I recall, Penn State was playing an away football game that weekend, thus there was potential for the weekend to be low key. Still, none of us wanted our weekend to go to waste. Each of us had already spent our weekly allowances on pizza and caffeine-laced drinks to help ease our Monday through Friday studying. That meant we were now forced to read the school newspaper in search of anything that was free.

As our luck would have it, there was indeed a free event coming to Penn State. The newspaper touted an international cultural event. It exclaimed that Taiwan was sending a goodwill circus to tour America, and one of its stops was to be on our campus. To the Taiwanese people, this was a chance to gain recognition as a separate entity from mainland China. To us, this was a chance to do something for—you guessed it—free!

The lights dimmed as we awaited the opening act. The lights flashed on, the music began, and there on stage was a man with ten sticks and a stack of plates. In a fit of choreography, he pranced across the stage and planted a stick on the floor, holding it firmly. One end touched the ground, and the other end, about six feet high, protruded into the air. Next, he took a plate, and began spinning it on top of the stick. I was not impressed until he walked away from it.

This one-man circus spent the next five minutes repeating his steps with the other sticks and additional plates. Eventually, he had ten sticks with ten spinning plates, all performing simultaneously. As one stick would start to tilt and the plate would wobble off its horizontal, he would react by running to the plate and increasing its spin. Our performer spent the next ten minutes leaping all around the stage, from one stick to the next, maintaining the spinning plates. The sweat on his face glistened in the theater lights. He ran faster and faster, but ultimately the plates wanted to fall, and he could not keep up the pace. He concluded his act by letting the plates fall one by one, gracefully into his arms, while allowing the sticks to fall to the ground. We cheered. It was free.

The Taiwanese circus performer had done a marvelous job entertaining us. He also did a marvelous job depicting a process that is always occurring around us. That process is called entropy.

Recall that entropy, by definition, implies that everything is tending toward decay and atrophy. In scientific terms, we state that matter is going from a high state to a low state. Certainly, once our performer grew weary from maintaining the spinning plates, each plate and stick gravitated toward a lower state. In a nutshell, the plates and sticks wanted to fall.

Entropy was thus in action as it sought to bring down the sticks and plates. Entropy is also in action wherever meteorology takes place. In fact, weather plays a supporting role in the entropy we experience here on our earth. Think about the degradation that occurs as a result of most types of weather. For example, due to rain, soil erodes and crops and topsoil are washed away. Rain also causes mudslides. These in turn, cause hills to tumble and homes to get swept away. In addition, rain can cause metals to rust. These are examples of entropy in action.

Consider when it is windy. The wind causes trees to blow over. This is entropy. Structures built by man often collapse in great windstorms. Again this is entropy. The winds whip up the waves and tides. The result is beaches that erode and coastal communities that flood. Once more, entropy is occurring.

The sun is also a source of entropy. Too much sun can burn up crops and dry up the landscape. Decay is thus taking place. As the sun heats rocks, they expand, crack, and disintegrate. The result is that matter is encouraged to go from a higher state to a lower state. Degradation wins out as entropy dominates.

Finally, anytime we note that something has the weathered look, we are really expressing acknowledgement of entropy that has occurred. We joyfully photograph weathered barns, when in reality we are documenting a point in its path toward demise. We talk of the character that a weathered fence brings to a property. Again, that fence is on the road to collapse. The weathered coastline that we flock to as tourists is no more than a disintegrating landform. Meteorologically induced entropy strikes wherever we glance.

BIBLICAL SOURCE OF LAW 7

The six previous laws of meteorology can all be traced back to the first few verses of Genesis. This is where they have their common Biblical roots. This is also where God initiated them. These first six laws all have something else in common, though. They are laws that deal with very positive aspects of weather. After all, they are laws that deal with the good that was imparted into the world by God, during His week of creation.

Our seventh law, however, is definitely one that is out of character when compared to the six before it. It is a law that has negative implications. It is not surprising that this law does not manifest itself, with regard to weather, until later in Genesis after creation. In fact, it is my belief that entropy came about only after the fall of man in the Garden of Eden. Entropy then became a consequence of the events in that garden. From the time of those events, entropy abounds throughout the Bible.

Meteorological examples of entropy in action can be cited in many locations in the Bible. The great flood in Noah's time stands out at the beginning. In Genesis 7:4 the Lord tells Noah that "Seven days from now I will send rain on the earth for forty days and forty nights, and I will wipe from the face of the earth every living creature I have made." God was going to use His weather as a tool for a quick destruction of life on earth. Forty days of rain meant rapid entropy was to take place.

The entropy caused by weather is revealed many more times in the Bible. We just noted what a forty-day flood could do to the planet. Earth's more typical rains are still a contributor to entropy as well. Job 14:19 states "...as water wears away stones and torrents wash away the soil...." Here, river water and rainwater are described as major sources of decay.

The converse of rain is sun. Even sun causes its share of degradation around the globe. In Psalm 90:5-6, it is written, "...the new grass of the morning—though in the morning it springs up new, by evening it is dry and withered." This concept of the sun's destruction is echoed again in James 1:11. Here the passage reads, "For the sun rises with scorching heat and withers the plant; its blossoms fall and its beauty is destroyed." The plant has gone from a high state to a lower state. The sun is the meteo-rological tool in this example for hastening entropy.

Hail is another aspect of weather that leads to de-struction, or entropy. The Bible contains several accounts of hail doing major damage. In Psalm 78:48, we note "He gave over their cattle to the hail...." Later in Isaiah 32:19, we learn that "...hail flattens the forest and the city is leveled completely...." Hail is definitely a source of entropy.

Severe storms often contain damaging winds. This damage is yet another example of entropy in action. In Psalm 103:16, we see that "...the wind blows over it and it is gone...." A higher state going to a lower state is por-trayed here. Isaiah 41:16 is a passage that reinforces this concept: "...the wind will pick them up, and a gale will blow them away." Wind results in entropy.

Here is a final Biblical example of entropy from me-teorology. This example involves the destructive power of lightning. Psalm 78:48 reads, "He gave over...their live-stock to bolts of lightning."

In reality, any type of weather you can imagine is in one way or another leading to the decay of the earth. In our current era we are vividly aware of this fact, due to worldwide media coverage of world weather calamities. As far back as the time of Noah, God was using weather as a tool to support entropy around the world. The writ-ers of the Bible documented this fact for us over and over again. They knew then, and we know now, that normal rain, floods, hail, wind, sun, lightning, and all other storms are major contributors to the decay of our planet. Weather hastens entropy.

BIBLICAL EXPANSION OF LAW 7

As you can imagine, weather is not the sole contribu-tor to entropy on our planet. In fact, entropy occurs at many other levels, even in respect to our own lives. Job cried out his frustration over this fact on more than one occasion. In Job 13:28, Job is coming to terms with his harsh realization that "...man wastes away like something rotten, like a garment eaten by moths." This is certainly a happy thought! It's true, though. Entropy strikes us all and takes down our earthly existence.

Job's frustration and cries carry on verse after verse. In Job 14:1-2, he states that: "Man born of woman is of few days and full of trouble. He springs up like a flower and withers away; like a fleeting shadow, he does not endure." Job is on a roll as over and over he laments entropy's successful ability to claim life.

Entropy is a force that does not discriminate. We know it not only attacks every living thing, but it also attacks the matter of our planet as well.

It is not just our lives, though, that succumb to entropy. Entropy is a force that does not discriminate. We know it not only attacks every living thing, but it also attacks the matter of our planet as well. Why not use Job one more time, as he notes in Job 14:18-19, "...a mountain erodes and crumbles...a rock is moved from its place...water wears away stones...torrents wash away the soil...." The matter of our planet is going from a high state to a lower state.

If you want to get really depressed, turn to the early verses of Ecclesiastes. Passage after passage declares the meaninglessness of life. Passage after passage declares the futility of everything. Later on, in Ecclesiastes 3:1-2, we read "There is a time for everything, and a season for every activity under heaven: a time to be born and a time to die...." Entropy is waiting to claim us. There's more, though. The concluding verses of Ecclesiastes 3 go on to truly express the hopeless view of life that so many people have had through the ages. Focus on Ecclesiastes 3:19-20, where it is written, "Everything is meaningless...to dust all return." Entropy conquers all, is the message. Entropy cannot be overcome, is also the message. But, is entropy the end of the road? Is it the conclusion and final act of life?

LAW 7: PATHWAY TO JESUS CHRIST

There is no doubt that all of us can relate to entropy in our lives. In fact, there is not one of us who is not touched and affected by entropy on a daily basis. Everything around us is tending to fall apart. Perhaps it's our health, or maybe it's the road we drive on to go to work.

Maybe it's our homes, or our social lives. Entropy is constantly at work around us.

Let us consider a typical day in my life. Yesterday is a wonderful example. My wife, Pam, and I woke up, and she hopped in the shower first. She mentioned as she got out of the shower, that mildew was making a comeback on the tiles and grout around the tub. OK, so my suburban Seattle house is not falling apart, but mildew formation is certainly one way that my house is going from a higher state to a lower state.

Next, my wife was falling behind schedule. Time was working against her. I believe time is a form of entropy that weakens, but I'll save that for another forum. Pam asked me to help her by helping to get our son, Austin, ready for school. She wanted me to pick out his clothes. As I grabbed a pair of socks from his dresser drawer, I couldn't help but notice that one of the socks had a hole in it. Entropy had struck again. Of course, I made him wear it anyway!

When it was time for breakfast, I served up an impressive bowl of cereal. Yes, I come from a long line of gourmet chefs. My son poured the milk on his cereal, took a bite, and quickly spit his mouthful back into the bowl. One whiff of the milk from the carton confirmed that the milk had gone bad. Entropy, entropy!

Time had arrived for me to drive Austin to school. As we walked to the car, I noted that one of my tires looked low. I mentally made a note that later I would fill it up with more air. I wasn't too concerned, though, as the tire was just about worn out. Its life span was about over. Yes, entropy affects tires.

Entropy also affects roads. On the way to Austin's school, the ride was anything but comfortable. I am used to air turbulence when flying, but ground turbulence is a different matter. Nonetheless, we hit ground turbulence as we drove on our town's uneven, badly patched, potholed streets.

We were within one block of Austin's school when we looked off to the right and saw a partially burned house. Surrounding the house was a barrier of yellow police tape. The house had been on fire last night, and today no one was allowed within fifty feet of the dwelling. Fire accelerates entropy.

Finally, we arrived at Austin's school and I walked him to his classroom. On a typical day, his teacher would be waiting at the classroom door to greet the children as they arrived. Instead, on this day, there was a strange woman. She was a substitute teacher. I inquired about Austin's regular teacher and was told that she would be back in about a week. The reason for her absence was that she had come down with a bad case of the flu. Entropy strikes again.

Was that a gray hair staring back at me? As I winced in reaction, I couldn't help but note the never-before-seen wrinkle that my grimacing face was exposing. You bet I let out a yelp as I recognized entropy attacking me on all fronts.

I don't typically go to work until afternoons or evenings. So after dropping Austin off at school, I headed home. I passed a truck that had a pseudo medical caricature on it. The truck was titled the VCR Medic. What's the world coming to if even our videocassette recorders fall apart? Thank goodness for the VCR Medic to combat the entropy that was striking at the heart of home entertainment!

When I got home, my wife was already gone. She is a physical therapist and generally works traditional 8-to-4:30 hours. Two hours after being at home, I got bored. So I did what every husband does, I called my wife. She picked up the phone, but let me know she couldn't talk right now because she was busy with a stroke patient. She asked me to stop by her facility at lunchtime, but I declined. I rarely visit the facility where my wife works, because the condition of most of her patients depresses me. She has stroke patients, motorcycle accident patients who are paralyzed, patients with memory loss, and so on. The health effects of entropy are very focused in Pam's place of work.

Eventually, it was time for me to get ready for work. I stepped up to the mirror and began combing my hair. Was that a gray hair staring back at me? As I winced in reaction, I couldn't help but note the never-before-seen wrinkle that my grimacing face was exposing. You bet I let out a yelp as I recognized entropy attacking me on all fronts.

Hastily, I grabbed my razor and started shaving. After a moment I realized that my razor blade had dulled from previous shaves. OK, it's one thing to have entropy give me gray hair and wrinkles, but leave my razor alone! I fumbled around attempting to discover where I last put my replacement blades.

Finally, I was ready. All I had left to do was put on my clothes for work, and I would be on my way. I took my pressed white shirt off its hanger from the dry cleaners. I slipped my arms through the sleeves and began buttoning the front. As I gingerly took hold of the third button up, it popped right off its thread. Thankfully, I had another shirt I could put on; I would have to deal with entropy's attack on my first shirt, later.

My drive to work was fairly uneventful. I did pass a plumbing repair truck, racing toward someone's home no doubt. Pipes were breaking down and a fix was urgently needed. Oh, entropy. Next, I passed a tow truck pulling a broken-down car. Entropy again. Driving on, I noted nine car repair facilities between my home and work. Each one was loaded with broken-down cars, just like the one that had passed me on the tow truck. The effects of entropy could not be missed. I scurried into work, hoping to escape entropy's long arm.

I arrived at my desk and began pondering stacks of weather maps and data. One item that caught my eye was a notice from the National Meteorological Center near Washington, D.C. The notice was sent to inform all meteorological users that their giant government computer had just stopped working, and they were having trouble figuring out why. The result would be a lack of future data and modeling information for quite some time. Could it possibly be true that entropy had struck a major computer near our nation's capital?

Now it was time for Plan B for making a forecast. I looked at a satellite loop from earlier in the day, and was struck by the preponderance of tropical moisture streaking toward Seattle from Hawaii. This kind of weather pattern we affectionately call the Pineapple Express. Hawaii is basically exporting its high liquid-content air toward us. The net result is that very heavy rains are likely in Western Washington.

Two hours went by and the skies opened up. Two hours beyond that and trouble began. The winds picked up, the rains came down, and damage reports began rolling in. We were in the midst of a severe storm. There would be no stopping entropy now. Our news crews were dispatched to the sites of three mudslides. Homes were sliding into Puget Sound. Wind was knocking down trees. Trees were taking down power lines and phone lines. Floodwaters were beginning to sweep through several valleys. All of this was making the degree of entropy that began my day look pretty good. Right now, entropy was becoming extreme.

I ran in and out of the studio, doing weather cut-ins one after another. After a long night, the weather started calming down. The Pineapple Express had "doled" itself out, and now the rains were beginning to ease into lighter windswept showers. Our main newsman was finally sent home and his replacement arrived. His replacement was a fine young lady who had a knack for telling it like it is. As she came in the door, everyone asked her how it was out there. Everyone wanted to know what she saw as she drove to work. I believe she let her priorities show. She responded to the crowd by taking her brush out of her purse, combing it through her locks to reestablish her fallen hair, and letting everyone know in no uncertain terms that it was a really bad hair day. I guess entropy is relative.

Entropy is certainly a normal characteristic of everyday life for us here on earth. I challenge you to spend a day contemplating the break down of our world. Whether you watch the news, or you watch the world as you drive across town, entropy is a force that will stand out. As you can imagine, the effects of entropy are not exactly wonderful.

As if normal entropy isn't bad enough, there are those out there who are out to hasten entropy in our society. They do this in hopes of making a quick buck. Consider the drug dealers and liquor dealers. They will sell you products that quicken the entropy of your bodies. Your brain cells disappear, your liver rots, your life goes into upheaval, and the effects emanate out to those whose lives you touch.

There are also those out there that are exploiting entropy, again in hopes of making a quick buck. They take advantage of entropy's existence, and peddle products that will supposedly reverse the entropy process. How many commercials have you seen that target men by advertising ways to get the gray out of your hair? How about those commercials that tout a product that will grow hair back in areas where you have gone bald?

Ladies, you are being targeted too. My favorite product that will fix your entropy is called night repair makeup. Yes, while you sleep, your face is repairing itself from the damage you noted the night before. Of course, if that doesn't work, then you can always purchase plain makeup to cover up your entropy.

Entropy, however, can strike beyond our own physical appearance. That's where other companies step in with their products. In my mailbox, I must find at least one ad per month from the Lawn Doctor. The promise is that this lawn service will change a disarrayed yard into something nice.

Other businesses are out to take advantage of entropy on the home front. How many ads have you seen for a $99 divorce? If entropy hits your marriage, there are those clamoring to take advantage of it by competing with their low-priced divorce packages. I'd much rather view an advertisement pushing a $99 marriage counseling package, but I have yet to see those.

Some entropy is merely an illusion, created by a manufacturer. Again, there is money to be made. My wife pointed out that women's shoe manufacturers often purposely declare a style as old, even though you may have purchased the style recently. You are encouraged to buy a new version of the shoes. In this case, entropy hasn't really happened, but our mindset is such that we accept the ploy. We are so accustomed to entropy in our lives that we believe it when we are told that our original shoes have fallen by the wayside and are out of fashion. We then dutifully buy a new pair of shoes.

This last example brings us to an important point: we must pick which forms of entropy we are going to allow ourselves to get caught up in. To put it another way, as in the case of our Taiwanese circus performer, we must choose which plates we want to spin. Otherwise, the potential exists to spend every waking moment combating the entropy in our lives. Ecclesiastes 4:6 deals with this concept. In fact, this passage almost sounds like a Chinese fortune cookie. It states, "Better one handful with tranquility than two handfuls with toil and chasing after the wind."

How many of us spend large portions of our precious time merely chasing after the wind? When I lived outside of Richmond, Virginia, I had a neighbor who spent his valuable time manicuring his yard. He was obsessed by it. His kids would be playing, but he was working on his yard. His wife would be trying to talk to him through the window, but he'd be cultivating his yard. We believe he actually lived in his yard, for he was always there. We slowly watched as this man's children started playing in other yards with other families. We also watched as this man's wife began taking long walks, hoping to socialize with neighbors who might give her a minute. The point I am making is that we need to spin plates that matter. This man's wife is a plate. This man's children are a plate. Unfortunately, the only plate that mattered to our neighbor was his yard. He spent his hours choosing to spin a plate that unfortunately reflected his priorities.

What priorities are you fixated upon? It's amazing how many folks are fixated on the entropy in their lives. You learn to readily recognize these people. You ask them how they are, and they are prepared to answer with a list of all the different kinds of entropy affecting them. Their response may include "I think I'm catching a cold." "My windshield wiper started smearing when it wipes." "My TV remote control died." "My dog has fleas." "I broke a nail!" These people are depressing. Sometimes they don't even realize that others are spinning plates as well. Instead they complain as if they are the only ones.

In actuality, entropy is combating us all. It is a major component of life. As I stated earlier in this chapter, I believe entropy was imparted into the world as the fall of man occurred in the Garden of Eden. After Adam and Eve used their free will and ate the forbidden fruit, the world has never been the same. Suffering and toil became part of life's equation. In Genesis 3:19, God declared to Adam and all that follow, "...to dust you will return." The ultimate entropy for us all, death, would now stare us down from the moment we are born.

As depressing as entropy sounds, there is actually a good side to it. Part of its goodness is that it motivates us to work and to problem solve. If entropy did not exist, then it's possible that life would be boring. Think about it. Most of us would not have jobs or a purpose in life. If entropy didn't attack our cars and appliances, we wouldn't need repairmen. In fact, everything we own would then last forever, so we wouldn't need manufacturers. If entropy didn't attack our bodies, we wouldn't need the medical profession. Most importantly though, if entropy didn't attack us, we wouldn't need or seek God.

Entropy leads us to Him. Initially some of us are quite adept at spinning our plates without God. Ultimately though, the final entropy, death, draws us toward God, unless we choose to spin that plate containing death on our own as well.

As we saw in the Old Testament, even entropy was not always enough of a lure to bring God's children back into obedience to Him. In fact, God's children often rebelled and turned away from Him. They spun their plates on their own or with false gods, and were overwhelmed by the heavy consequences of their actions.

Entropy is a result of our sins, but the cure is a result of our Lord and His love for us. His cure for us is a permanent infusion of Jesus Christ into our lives. The downhill slide in our lives can be reversed!

There is a cure though, for all of us, who are overwhelmed by having to spin plates and deal with entropy. Entropy is a result of our sins, but the cure is a result of our Lord and His love for us. His cure for us is a permanent infusion of Jesus Christ into our lives. The downhill slide in our lives can be reversed!

Jesus Christ can bring us renewal and eternal life. He showed us the way, by travelling that path first. He died, but then He rose. Our lives, with Christ, can exactly mirror this miracle. We don't need to be fallen. We now have the tool to be risen!

Entropy means everything is wearing down. Entropy means everything is futile. Entropy is no doubt the devil's playground, for as entropy weakens us, Satan surely steps in. You have a choice, though. You can choose not to let Satan step in. Your life's path doesn't have to be down-hill. It is as simple as displacing the devil with Jesus Christ.

As always, our free will comes into play, and we must make a choice. Our lives can simply be a downward spi-ral. Entropy can always dominate. Death can be the final entropy in our lives. It can be the end of our story. That is all one choice.

However, the alternative is a much more enticing choice. Choosing Jesus Christ means entropy doesn't get the better of us. It means our burden is eased. It means we have spiritual support on earth as we spin our plates away. It means our priorities change and we cease chas-ing the wind by spinning meaningless plates. It means we have a defense as the devil tries to bring us down.

Choosing Christ to give us an upward direction is the wisest thing we can ever do. Proverbs 15:24 says, "The path of life leads upward for the wise to keep him from going down to the grave." Choose your preferred direc-tion, or entropy will decide for you.

NINE
......●●●●●●●●●●...●

Choosing to Walk Down God's Path

My deepest prayer is that you choose to make the Lord of the skies, the Lord of your life. That is the soul reason I wrote this book. Every day, I stand before a television camera, and it saddens and frustrates me to realize that so many people make my job my identity. It sickens me that so many folks look up to me only because of what I do for a living. Television is not who I am. This book, and what it says, and what it stands for, is who I am. My identity, purpose, and self worth are derived from my relationship with Jesus Christ. It took quite a path, though, to get me to this point.

Weather has been that path, leading me straight to God. From

> **My identity, purpose, and self worth are derived from my relationship with Jesus Christ.**

the moment I made the connection that weather was His signature in the sky, that path began evolving. Over time it turned into a two-lane road, and now it's an eight-lane highway. As my understanding about God and His love for me matures, I find myself riding in the express lane of that highway. I echo Psalm 19:1, "The heavens declare the glory of God; the skies proclaim the work of his hands." I can't look at anything meteorological anymore without being struck by the realization that our Father, God, is the Lord of the skies, and the Lord of everything else!

I also can't look at weather anymore without recognizing the parallels between the science of meteorology and the plan God has for us. It is comical that before my journey of faith began, I really believed I personally had discovered the Seven Laws of Meteorology. How humbling it was for me when I discovered that those seven laws were initiated by God and recorded in the pages of the Bible. How exhilarating it is that these seven laws reveal so much about God's character and His plan for us, and that we are free to choose and embrace it.

The Seven Laws of Meteorology have led me to an acute awareness of God. So it is only natural that I see God's fingerprints as I look up into the atmosphere. Most likely your profession is something other than the field of meteorology. I hope you realize that your field can also initiate a path leading you to God.

Perhaps you are a geologist. Your rock-studded path to God is literally right in front of you. I can only imagine the exciting way that God's fingerprints must emanate from the minerals and rock types that make up the story of earth. The choice is yours if you want to see God in your work or not. God, though, is Lord of the rocks!

Perhaps you work with lumber that has been cut from the magnificent trees of our world. You then must certainly recognize the fine beauty in a good piece of wood. God's handiwork is before you, only you must choose to see it. You have the privilege of adding your workmanship to God's workmanship as you work the wood. That opens up a path that again leads toward Him. God is Lord of the trees!

Perhaps chemistry is your profession. The chemist's favorite chart, the periodic table, reflects the elements that compose our world. The periodic table brings a sense of order to the chemical tools handed us. They are building blocks given to man by the Almighty. Chemistry is a system just like weather is a system. God's fingerprints are surely manifest. You must only choose to recognize that. Your path to God awaits you. God is Lord of chemistry!

Perhaps you are a musician. Like chemistry, music is a system. It is organized and has the ability to stir up emotions. It is an expression of our humanity, and with the proper attitude, it can bring plenty of glory to God. Choose to let God come forth from your music. You will then be headed down that path to God, Himself. God is Lord of music!

Perhaps you are employed in the medical profession. You are aware that the human body and its workings are absolutely intricate. Such intricate design demands an ingenious creator. The body is nothing short of miraculous. God is indeed before you. Choosing to recognize this will place a pathway before you, leading to Him. God is Lord of the medical wonder known as our bodies!

Perhaps you have a career that is intensely people interactive. You might be a social worker, a lawyer, a television news reporter, or a bank teller. Your search for God's fingerprints is as close as the people you deal with every day. You become a witness to the choices people make as they utilize the gift of free will that God gave them. With their free will, the terrible side of folks often comes out. However, with free will, the goodness has just as much potential to shine. The choice is theirs, but when people choose to reflect goodness, you get to witness God at work.

After all, He does some of His finest work through people. Recognizing that fact can put a whole new light on the way you view people. Choosing to recognize that God is inside people can beckon you to your own path leading to Him. God is Lord of people!

Now it is your turn. You have the ability to take that first step down a divine path that God has personally laid out just for you. As you travel that path, you will no longer see God as an abstract figure off in the distance. He will no longer be a character you merely have heard or read about. Instead you will rediscover the Lord and come to know a totally real, and infinitely loving Father that has only the very best in mind for you.

God's best for you is reflected by the seven laws noted throughout this book. Recall Law 1, where you learned that the sun takes center stage in weather. You then saw how God is in the habit of centering more than just the weather. In fact, He gave you His son, Jesus Christ, to be the center of your life. He is giving you a loving, stable, merciful, positive, and life-changing power on which to anchor your life. He wants you to rediscover Jesus Christ and the infinite value that a relationship with Him can bring to your life.

Consider Law 2, where you focused on weather being a total system. God designed lots of systems, not just weather. His most grand system was the system of Jesus Christ. So that there would be no doubt regarding Christ's claim as your Messiah, God graciously foretold in the Old Testament all that you would need to know to recognize Him. It was foretold about His birth, His life, His death, and His resurrection. The New Testament, history, and millions of transformed lives uphold Jesus Christ's true identity as the Son of God. You now have the opportunity to be a part of the grandest system ever designed by the Lord Almighty. By accepting Christ as your savior, you will never doubt again what your place is in this world. You will have unwavering knowledge of how you fit in. You will know your immense value.

Law 3 devoted itself to the way that weather is always in the process of smoothing out excesses and deficits. You learned that in weather, excesses tend to migrate toward regions of deficit. You also learned that in your life, there are a lot of excesses that need to be diffused. One of these excesses is the sin that you produce. In God's infinite wisdom, He gave you a vehicle for releasing that excess sin. Jesus Christ is that vehicle. Your sins are transferred to Him who is unblemished. You can then be free of the bondage of sin. Once free, you are equipped to offer the greatest message of all, the message of God and His Son. The world is absolutely desperate for such a message.

Law 4 certainly reflects some of the best God has for you. An attitude of "whatever it takes" is behind the design of weather. So many peculiar components combine to make up the system of weather. There is that cockeyed tilt of our earth. There is that spin of our tilted earth. There is that rotation around the sun of our spinning, tilted earth. Each element is "whatever it takes" to help make the whole system work.

God is the mastermind behind these workings of weather. Likewise, God is the mastermind behind getting His children delivered from unending cycles of sin. God did "whatever it takes" as He made a supreme sacrifice on our behalf to bridge the gap between Him and us. He sent us Jesus Christ. Jesus now stands ready to enter into your heart, and connect you to the Father. You then get the privilege of experiencing and fully grasping the miraculous ways that God applies His "whatever it takes" attitude in your own life. It's an experience that is second to none!

Balance in weather was the theme of Law 5. You discovered that couplets-of-opposites exist in not only meteorology, but also in just about every other facet of life you encounter. Balance is a trademark of the world God made around you. This in turn implies that you ceaselessly have to make choices. To make these choices, you must draw upon the gift of free will that the Father gave you when He designed you.

Making choices is a mighty responsibility and often a burdensome one as well. Forces of evil are always nudging at you, trying to influence the choices you make. God knew that, and gave you a tool more powerful than any force of evil out there. God gave you Jesus Christ. He can be a shaper of your life, and with Him, you are able to combat the evil forces that try to take a hold of your life. Take that path God has laid out for you and never again will you have to fight evil alone. You become a player on the side that will ultimately win.

Law 6 focused on the purpose that weather serves. You saw that because weather is really of God, it must serve a purpose that is positive. After all, that is His character. Weather, and all of God's other creations, reflect this positive purpose, and thus Him. Taking it all a step further, you are a creation of God, just like weather. Therefore, you are also intended for positive purpose. Unfortunately, your free will often gets in the way of that positive purpose being served.

Thankfully, though, God gave you His Son. By accepting Jesus Christ into your life, you become equipped to turn your life around. Now you may indeed fulfill that positive purpose for which you are intended. That positive purpose will then literally become the focus of your life. No longer will your life seem meaningless and without purpose. Life is definitely not futile. There is a reason for you. Choose to team up with God and His Son, and your purpose will be revealed.

The last Law of Meteorology, Law 7, referred to the way that weather aids in carrying out the process of entropy here on earth. The planet is breaking down, but it is often a result of more than just weather. In fact, it is typically a consequence of mankind's own actions. In any event, entropy can be a depressing topic. It means you come face to face with the realization that everything is wearing down, including yourself. On your own, you can only pathetically strive to counter the forces of entropy, including death. Of course, it is to no avail. Entropy could easily be the end of the road for you.

The loving Father, though, mercifully gave you the one and only way for truly conquering the entropy in your life. That way is Jesus Christ. The path to Him will not succumb to entropy. When the path seemingly ends at the feet of God and His Son, it's really just the beginning. Travel the path to the Almighty, and both abundant and eternal life will lie before you.

Until now you've heard about God. You've heard about Jesus Christ. Maybe you even have a relationship with the Holy Ones. Your relationship, though, can be so much more rich and fulfilling. It is time to take a walk down that divinely prepared pathway that lies before you. Down the trail, you will rediscover God, His Son, and yourself.